T0178112

The cause of global change has been the subject of heated debate in the past few years, especially in relation to climate change and biodiversity decline. However, a systematic explanation for changes in the biosphere at global level has still to be found. In this volume, a wide range of viewpoints from ecology and economics are surveyed to see if some light can be shed on this problem. Economists analyse how economic growth predictably alters the earth, and ecologists consider how the drive for fitness and consequent population growth changes the globe. Both look at the institutional interface between humans and the biosphere, and explain global change as the consequence of human non-cooperation and conflict. The conclusion is left to the reader: the object of this volume is to initiate debate on global change at this most fundamental level.

THE ECONOMICS AND ECOLOGY OF BIODIVERSITY DECLINE:
THE FORCES DRIVING GLOBAL CHANGE

THE ECONOMICS AND ECOLOGY OF BIODIVERSITY DECLINE: THE FORCES DRIVING GLOBAL CHANGE

Edited by

TIMOTHY M. SWANSON

*Faculty of Economics, Cambridge University
and Director of Biodiversity Programme, CSERGE – University
of East Anglia and University College London*

CAMBRIDGE
UNIVERSITY PRESS

CAMBRIDGE
UNIVERSITY PRESS

University Printing House, Cambridge CB2 8BS, United Kingdom

One Liberty Plaza, 20th Floor, New York, NY 10006, USA

477 Williamstown Road, Port Melbourne, VIC 3207, Australia

4843/24, 2nd Floor, Ansari Road, Daryaganj, Delhi - 110002, India

79 Anson Road, #06-04/06, Singapore 079906

Cambridge University Press is part of the University of Cambridge.

It furthers the University's mission by disseminating knowledge in the pursuit of
education, learning and research at the highest international levels of excellence.

www.cambridge.org
Information on this title: www.cambridge.org/9780521635790

First published 1995
First paperback edition 1998

A catalogue record for this publication is available from the British Library

Library of Congress Cataloging in Publication data
The economics and ecology of biodiversity decline : the forces driving
global change / edited by Timothy M. swanson.
    p. cm.
Papers originally presented at a symposium held at King's College,
Cambridge, in July 1993.
    ISBN 0 521 48230 5 (hc)
    1. Biological diversity conservation - Congresses.   2. Economic
development - Environmental aspects - Congresses   3. Biological
diversity conservation - Economic aspects - Congresses.   I. Swanson,
Timothy M.
QH75.AlE27   1995
333.95 - dc20   94-42045 CIP

ISBN   978-0-521-48230-1   Hardback
ISBN   978-0-521-63579-0   Paperback

Contents

Contributors

Professor Gardner Brown
Department of Economics
University of Washington
Seattle, WA 98195, USA

Professor Colin Clark
Department of Mathematics
University of British Columbia
Vancouver, Canada BC V6T 1W5

Dr Gretchen C. Daily
Energy and Resources Group
Building T4, Room 100
University of California
Berkeley CA 94720, USA

Professor Paul R. Ehrlich
Center for Conservation Biology
Department of Biological Sciences
Stanford University
Stanford CA 94305, USA

Professor Madhav Gadgil
Centre for Ecological Science
The Indian Institute of Science
Bangalore 560 012, India

Professor John M. Hartwick
Department of Economics
Queen's University
Kingston, Ontario, Canada K7L 3N6

Professor Jeffrey A. Krautkraemer
Department of Economics
Washington State University
Pullman WA 99164-4860, USA

Dr Norman Myers
Upper Meadow
Old Road
Headington
Oxford OX3 8SZ, UK

Professor Charles Perrings
Department of Environmental
  Economics and Environmental
  Management
University of York
Heslington
York YO1 5DD, UK

Professor Jonathan Roughgarden
Department of Biology
Stanford University
Stanford CA 94305, USA

Professor Bob Rowthorn
Faculty of Economics and Politics
University of Cambridge
Sidgwick Avenue
Cambridge CB3 9DD, UK

Professor Roger Sedjo
Resources for the Future
1616 P Street NW
Washington DC 20036, USA

Dr R. David Simpson
Resources for the Future
1616 P Street NW
Washington DC 20036, USA

Fraser D. M. Smith
Center for Conservation Biology
Department of Biological Sciences
Stanford University
Stanford CA 94305-5020, USA

Dr Douglas Southgate
Department of Agricultural
  Economics and Rural Sociology
Ohio State University
2120 Fyffe Rode
Columbus OH 43210-1099, USA

Dr Timothy M. Swanson
Faculty of Economics and Politics
Sidgwick Avenue
University of Cambridge
Cambridge CB3 9DD, UK

Preface

Why is the world losing its biological diversity? And why, if the estimates are to be believed, are we losing it at an unprecedented rate? While most people have their individual theories – population, developing country poverty, misdirected overseas aid policies, etc. – rigorous appraisals of the phenomenon are few and far between. Yet, if we do not understand causes, the design of policy is fruitless. CSERGE is Europe's premier institution investigating the forces for global environmental change and the design of policy to correct the undesirable aspects of that change. CSERGE's directors convened a workshop in the splendour of the surroundings of King's College, Cambridge, and brought together the world's leading experts on the issue of causation. The meeting was biased towards economists, for a good reason: the science of environmental economics has taught us that, if we want to understand environmental change, we have first to investigate the workings and, more importantly, the 'failures' of the economic system. But equally valuable insights can be gleaned from ecologists and anthropologists, sociologists and planners.

CSERGE entrusted the masterminding of the symposium to Tim Swanson, who directs CSERGE's biodiversity programme and who has himself produced some hard rethinking of the conventional wisdom in biodiversity management. In essence, the issue is one of incentives: incentives to destroy biodiversity through the conversion of land from high diversity characteristics to uses with low diversity characteristics; and incentives to invest in biological resources as a source of human wellbeing. Traditional conservation policies are slowly recognising that, without attention to incentives, there is little prospect for conservation. The 'fence it, police it' approach still has some role to play, but against the forces driving biodiversity loss this philosophy is increasingly outmoded.

We hope that this volume will contribute to the continuing rethink and the Directors of CSERGE thank Tim Swanson for convening the symposium and editing its proceedings.

David Pearce
R. Kerry Turner
Timothy O'Riordan

Directors, CSERGE

Acknowledgement

This volume reports the findings of a symposium held at King's College, Cambridge, in July 1993 on the causes of the loss of the world's biological diversity. The symposium was organised by the Centre for Social and Economic Research on the Global Environment (CSERGE), which is based at University College London and the University of East Anglia in the UK. CSERGE is a designated research centre of the UK Economic and Social Research Centre (ESRC) and the Directors of CSERGE express their continuing thanks to the ESRC for support for the Centre. The symposium also received financial support from the United Nations Environment Programme (UNEP), Nairobi; the UK Department of the Environment; and the UK Overseas Development Administration (ODA). The World Bank generously supported the travel costs of some participants. CSERGE thanks these sponsors for their positive and helpful response to the idea of the seminar. Finally, Miranda Roddy and Janet Roddy bore the burden of organising the seminar with admirable efficiency and good grace.

Cambridge, London, Norwich,   *Tim Swanson*
February 1995   *David Pearce*
                                     R. Kerry Turner

1

Why does biodiversity decline?
The analysis of forces for global change

TIMOTHY M. SWANSON

Introduction: why is this an interesting question?

Why is biological diversity in decline? To many, the answer to this question is obvious: the problem is rooted in the expansion of human society – the human population and the human economy. For observers of this persuasion, the solution is just as obvious: human societies must be deterred from continued expansion in order to avoid biodiversity depletion.

Other observers, however, reach the opposite conclusion. They see that much of the deforestation that is occurring and the degradation that is its consequence are the result of the underdevelopment of certain segments of human society. Poverty shifts certain peoples onto the frontier, and the lack of developed institutions results in degradation and the decline of the diversity resident in these regions. Since most biological diversity resides on the fringes of human society, as do most poor peoples, it is the lack of human development within those segments of society that then results in most biodiversity decline.

Between these two polar perspectives on the problem, there are intermediate possibilities. Diversity decline may be the result not of general human development and expansion but rather of specific choices of path of human development. For example, it is possible that human development could be made consistent with greater biodiversity, either through the more intensive use of a smaller area of land or by the more extensive use of a wider range of species. Biodiversity decline from this perspective is not the consequence of either development or underdevelopment but rather the result of a choice of a particular path of development.

This volume examines these issues from these various perspectives. It does not reach one determinate conclusion. This is because the volume is intended to serve the purpose of illustrating and illuminating the general analysis of the problems of global change. Appropriate policy making in those areas will be built upon reasoned analysis and debate of the fundamental causes of global problems, rather than 'knee-jerk' responses based upon individual intuitions. This volume is intended to illuminate the hidden complexities within a global problem such as diversity decline, and it is also hoped that it will illustrate the initial steps that are required in order to develop debate and ultimately to form consensus. For these reasons we place before the scientific and conservation communities this variety of perspectives on diversity decline, and allow these view points to argue for themselves.

In the remainder of this introductory chapter the thread of the various arguments is set forth and the organisation of the volume explained. Although there is considerable overlap between categories, the volume is none the less divided into four distinct parts, in accordance with the general nature of the arguments. The first part of the book examines the 'development path approach' to biodiversity decline. It is a collection of three pieces arguing that human society has a choice in regard to the amount of diversity that will be retained along its development path, and that this choice has thus far been made in a haphazard fashion, resulting in unmanaged diversity depletion.

The second part of the book examines diversity decline as a specific form of 'institutional failure': the failure to create institutions that internalise the values of biodiversity within the decision-making of states and individuals making conversion decisions. This is an instance where it is the lack of development of specific forms of human management institutions that is the root cause of biodiversity decline.

The third part of the book echoes this fundamental conclusion but points to the development of 'inappropriate policies' rather than the absence of institutions as a positive determinant of biodiversity decline. Policies exist that force the poverty-stricken segments of human societies onto the lands at the fringe of these societies, and into competition with all of the other species that must exist there; they are inappropriate for both the people and the resources with which they live.

The fourth part of the volume contains three pieces that see general human development as the causal force behind diversity decline. The thesis of the 'development-induced decline' school of thought is straightforward. The human species pursues development strategies rooted in self-interest and in fundamental conflict with other species' interests. Thus, there is little scope for marginal changes in human behaviour or institutions to affect this fundamental force for decline. Only the disavowal of expansionist human development policies has the potential to be effective for the conservation of biodiversity, from this perspective.

The remainder of this chapter outlines these various arguments for the benefit of the readership that is most time-constrained. For those with sufficient interest and adequate time, I suggest that you now turn to Chapter 2, and allow the various authors to speak for themselves.

Diversity decline from choice of development path

Within the economic framework, the conversion of natural environments presents a choice regarding the portfolio of assets that human society will retain. Assets are forms of capital (human, physical, natural) that generate a flow of services, and natural environments (the sort that contain biodiversity) constitute one sort of natural asset. One way to define the human development process is the rearrangement of the initial portfolio of assets in order to generate an alternative human-preferred flow of goods and services. This occurs by means of the conversion of one form of asset into another. Changes in the relative amounts of stocks of assets generate the desired growth in the flow of the human-preferred goods and services: human development.

For example, development may occur when a form of natural capital is converted into human capital, by the 'mining' of a forest for investment in a society's educational system. It may also occur when one form of natural asset is converted into another, e.g. the conversion of a forest into agricultural lands. These conversion decisions constitute human development if they generate a human-preferred flow of goods and services.

From this perspective the problem of biodiversity decline lies in the bias within existing development paths towards a reduced variety of 'biological assets' (i.e. species). This might result from a number of different factors within the human development process. The authors in the first section of this volume discuss how societies select development paths, and the implications of such choices for biological diversity.

Krautkraemer (Chapter 2) outlines these factors and discusses the general propensity for land conversion within the (current) human development process. In aggregate, he finds little in the way of a global bias towards natural habitat conversion, but there remains evidence of its specific application where this development strategy remains available – in those developing countries with tropical rainforest. The issue that is crucial to the conservation of biological diversity is: what factors will determine when human development will discontinue the conversion of natural environments, leaving some residual of diversity to yield services for future generations?

One important factor is the discount rate. Most directly, the discount rate is relevant because it is determined in part by the return on physical capital, and natural capital is converted when it is unable to meet the rate of competing capital goods. There is a substitution effect between different forms of capital. Therefore, a relatively greater return available on other capital goods means a lower stock of capital retained in the form of natural assets.

This is not the only effect from the choice of discount rate. Rowthorn and Brown (Chapter 3) argue that there are both substitution and output effects from discount rates. Low discount rates call for high values to be placed on future flows of goods and services, and hence call for relatively greater levels of investment in the present. If the conversion of land is one such form of investment, then it is possible that a *lower* discount rate will generate a greater stock of converted lands.

The point being made here is simply that human societies have a choice of development path determined by the manipulation of certain fundamental variables such as the discount rate. The mix of goods and services and hence the variety of assets are choices taken by societies of their own free will.

A final point in this regard is that sometimes an arbitrary choice (in the sense that it was a 'free choice' made in isolation by some human society) will determine the choices made and paths taken by many subsequent societies. Swanson (Chapter 4) makes this point in the context of biodiversity. He argues that the degree of conversion witnessed in developing societies is predetermined by the conversion decisions made by the first-developing societies. These societies selected a set of locally available natural assets around which to develop, but many subsequent asset selections have taken their shape in response to these initial decisions. Now societies that are 'catching up' attempt to leapfrog intermediate stages of development by means of transposing the selections made by previous developers to their own territories. In this way development is biased toward the conversion of natural environments to the same set of assets across the globe. This is diversity decline as the result of the uniformity of the development process across heterogeneous states.

In Part A, the decline of diversity is not seen as a foregone conclusion resulting from the progress of development across the globe. The mix of assets on which human society will rely is an important choice variable and if humans value diversity in the biological sphere, then it is possible to alter this mix by working through various fundamental policy instruments such as the discount rate. It is simply necessary to make intelligent use of the human capability to choose its own path of development.

Diversity decline as a result of institutional failure

Why do not the goods and services deriving from biological diversity figure more prominently in the decision-making process concerning society's portfolio of assets? If these assets and the flows they generate were accurately valued, then there would be little reason to be concerned about the incorrect choice of a development path. The chapters within Part B of this book argue that there are external values of biodiversity that do not enter into the calculations concerning asset selection. The failure of institutions to incorporate these values within the decision-making process is then the root cause of biodiversity's decline.

Hartwick (Chapter 5) argues that there are ecological values of bio-diversity that must be incorporated within the human estimation of asset flows in order for asset selection to operate efficiently. If the continued conversion of natural environments introduces some risk of discontinuous decline within human production systems (e.g. due to an 'ecological catastrophe' such as crop failure), then the value of such a hazard should be incorporated into the natural accounts. Otherwise, the failure of human institutions to value this hazard will lead to excessive conversions of natural environments.

Is it true that 'ecological catastrophe' may result from the conversion of natural lands and the consequent loss of diversity? Perrings (Chapter 6) shows that many ecologists believe this to be the case. This is because ecological systems represent the equilibrium deriving from the interface between a set of ecological functions. These functions depend in turn on the species that perform them. Systemic stability thus depends on the web of species of which it is constituted. The loss of a single species that performs one of these functions may cause the entire system to move toward a new state of the equilibrium. If the natural equilibrium is fragile (as is often the case) then a barely perceptible change in the initial conditions may generate a dramatic change in the equilibrium system. Hence ecologists view the possibility of discontinuous decline in the productivity of an ecological system as being a possible consequence of the loss of diversity.

Can institutions be reformed to value the whole of a system as more than the sum of its parts? Perrings views many if not most of the institutional failures regarding the valuation of diversity's services as local rather than global ones. Ecosystems tend to be localised and thus the internalisation of the externalities associated with diversity decline may be achieved largely through the reform of national institutions. The important ecological problems of biodiversity decline may then be addressed by means of ensuring that local communities are free to manage their own resources.

The chapters by Hartwick and Perrings are concerned with the insurance value of biodiversity, whereas Sedjo and Simpson (Chapter 7) analyse the same institutional problem in the context of biodiversity's informational value. They look very specifically at the complexity and costliness implicit in the creation of an institution based on property rights in naturally generated information. The bias in this instance is caused by the existence of property right institutions for the appropriation of the value of information emanating from human capital ('intellectual property rights'), and the lack of any analogous institution capturing the value of information deriving from natural capital. Since diversity constitutes information, and this information has been of demonstrated usefulness for human societies for thousands of years, the absence of an 'institution' that rewards diversity's informational potential constitutes a serious institutional deficiency.

Therefore Part B looks to find the bias toward the conversion of natural lands in the relative appropriability of the flows from different types of asset (human, physical and natural). The chapters in this section of the volume find that this bias emanates from the failure of institutions to take into account the insurance and informational values of diversity. Until these institutional reforms occur, there will remain a bias for the conversion of diverse forms of natural assets.

Diversity decline as a result of policy failure

It is not only the case that the undervaluation of diverse resources will cause their decline; the overvaluation of converted assets will generate the same result. Governments have long encouraged the conversion of natural assets because they have confused a side effect of previous countries' development processes with the actual process of development itself. Hence many developing countries encourage conversions directly by means of agricultural subsidies, land grants and other such policies.

As mentioned above, these policies are based on observations that confuse cause with effect. Since many countries have previously engaged in development that resulted in the conversion of their natural forests, currently developing countries view this as one of the possible 'causes' of development and hence encourage it. It is in this sense that these policies are 'failures': they attempt to 'force' development down misconceived paths.

Many of these countries also use these lands as a means of addressing two development problems simultaneously, solving neither satisfactorily. Undeveloped lands are left to the poorest segments of society, in the hope that both peoples and lands will acquire some of the benefits from development elsewhere. Hence those peoples with virtually no assets are left to those lands with the least management. Myers (Chapter 10) terms this the problem of the 'shifted cultivator' and finds that it underlies the vast majority of the depletion of the world's biological diversity. Forcing the fringes of society into fragile ecosystems is a recipe for human and environmental degradation.

Southgate (Chapter 8) also argues that it is the inefficient use of the extensive margins of society that causes biodiversity depletion. He finds that governments in developing countries often refuse to invest efficiently in the intensive development of inframarginal lands whereas marginal lands remain available for conversion, even if such conversion is demonstrably inefficient. It is an unwillingness explicitly to spend funds (albeit efficiently) so long as implicit expenditures (of natural assets) are still available. Governments continue to observe policies that treat these assets as wastelands rather than investments.

Gadgil (Chapter 9) puts these issues into broader perspective. He sees the problem of the 'shifted cultivator' as the outcome of a global contest for resources. As societies across the world develop unevenly, there are some segments that come to draw upon a wider resource base earlier than do others. As their needs grow they displace some communities from the resources and relationships on which the latter have depended. These displaced peoples become 'ecological refugees', not only in the sense that they have lost contact with a system in which they were effectively integrated but also because they are now integrated within a system they do not understand. This is destructive for both the people and their new environment. Ultimately it is the lack of concern for the effective development both of these peoples and of these lands that drives diversity decline.

Therefore the chapters in Part C make a strong case for the wilful ignorance of human society as a fundamental force underlying the decline of biodiversity. Policies are adopted that induce conversions in a misguided attempt to 'force' development. Other times the diverse services of these lands are lost by reason of their induced degradation, by means of simply shifting social problems between locations. In this case the motivation is not a misguided pursuit of development, but a clear lack of concern for both the people and the resource.

Diversity decline as the result of development

The most straightforward explanation for the decline of biological diversity is that it is necessarily depleted by the expansion of human populations and economies. Such a perspective is closely aligned with the biological view of the world – competing life forms battling for niches within a given resource constraint. The expansionist tendencies of one species must necessarily have negative implications for others and (the ecologists' addendum) potentially for the entire system. The chapters in Part D develop this theme.

Smith, Daily and Ehrlich (Chapter 11) stress the impact of pure scale of human populations on biodiversity. Given current rates of mortality and the age distribution across human populations, the global population will approximately double over the next 60 years (even assuming that fertility rates continue to decline). This means that the global population will have increased from approximately 3 billion in 1970 to 5.2 billion in 1990 and 10 billion in 2060 – a more than three-fold increase in the space of 90 years. Such expansion of one species' population must necessarily have ramifications for others competing for the same resources. Hence we can see that the three direct causes of most extinctions (overexploitation, habitat destruction, human translocation of species) result from this inherent conflict between humans and other species.

Clark (Chapter 12) emphasises the perverse incentives for human economic exploitation of environmental systems in general. As human systems expand they require even more resources to service them. The unpriced and underappreciated environmental sector is exploited ever more intensively as a way of meeting these increased resource requirements without paying a price for them. It is this 'negative feedback effect' resulting from increasing resource scarcity that hastens the approach of potential catastrophe. It is as if societies had an in-built mechanism for rushing toward the nearest point of potential systemic collapse.

The final word in the volume goes to Roughgarden (Chapter 13), who uses it to declaim the level of abstraction incorporated within the economic approach to biodiversity decline. He sees little usefulness to an approach that continues to incorporate patently silly assumptions (such as the unlimited substitutability incorporated within the Cobb–Douglas production function) or to elide crucially important features in the analysis of an environmental problem (such as an ecological system's dynamics). Economists cannot bias their models in such a fashion and then claim to perform objective analysis. Equally they should not criticise other disciplines for being forthright about their subjectivity while the bias within economics remains so deeply buried in the name of abstraction. Economists must gain some level of appreciation for the environment, individually and intellectually, if they are to analyse it responsibly.

Conclusion

The object of this volume is to instill a well-focused debate on the fundamental nature of the problem of global change. Many times a problem such as biodiversity depletion is assumed to have a straightforward cause and an easy-to-derive solution. When people jump to immediate conclusions based on such intuitions, the variety of conclusions reached can make consensus difficult to build. The contributors to this volume were asked to consider the causes of biodiversity decline and then to attend a small symposium (held in King's College, Cambridge, July 1993) to discuss these causes. Then these authors were asked to submit chapters representative of the viewpoints they raised at the symposium. This collection is the result.

The variety of perspectives represented here is indicative of the complexity of these problems. The commencement of this debate remains a first, necessary step toward the problem's solution. The ultimately correct policies will be derived from the integration of these various perspectives (and other competing explanations) into a common framework.

Part A

Diversity decline as choice of development path

2

Incentives, development and population: a growth-theoretic perspective

JEFFREY A. KRAUTKRAEMER

This chapter takes the view that biodiversity conservation is closely related to preserving natural environments. This viewpoint is based on the following: (a) biodiversity is the outcome of an evolutionary process and natural environments are critical for this process; (b) estimates of the rate of species loss generally are based on the rate of habitat loss (Wilson, 1992); and (c) biodiversity conservation fits nicely into Krutilla's (1967) conceptual framework of the economics of natural environments. It is difficult to pin down exactly what is meant by biodiversity or exactly how one would measure it.[1] Conserving biodiversity is not only about saving specific species or even specific habitats but, in some sense, it is about saving a diverse set of places that allows evolution to continue in a less human-directed way. Characteristics of biodiversity that match up with characteristics identified by Krutilla (1967) as important issues regarding preserving natural environments include the public good aspect of many benefits of biodiversity, the lack of close substitutes for biodiversity, the uncertainty about future availability and relative value of biodiversity, and the irreversible nature of biodiversity loss.

Forces driving biodiversity loss

Biodiversity loss comes primarily from conversion and degradation of natural habitats, although overharvest is responsible for the loss of some specific species. The focus here is mainly, if not exclusively, on habitat conversion; to some degree, the case of degradation is essentially the same in that it is a slower form of conversion. Conversion and degradation occur

---

[1] Indeed, it may not be possible to adequately measure biodiversity with a single index. See Vane-Wright *et al.* (1991), Solow *et al.* (1993), and Weitzman (1992) for measures of biodiversity.

in a variety of ways. There can be complete physical make-over such as clear cutting (clear felling) a tract of forest land or draining a wetland. This make-over can be irreversible if the microclimate and/or soil processes are altered in significant ways. Degradation or conversion of natural habitats can occur when too much of an important constituent of an ecosystem is withdrawn – for example, the use of irrigation water in a river basin.[2] Alternatively, degradation or conversion can occur if too much of some material, natural or synthetic, is added to the ecosystem, including the introduction of exotic species or pollution from pesticides or other foreign materials.

Biodiversity provides a variety of services to the human community.[3] Given that biodiversity is valuable, then why do conversion and degradation of natural environments occur? This chapter uses a growth-theoretic perspective to organize a discussion of the forces driving biodiversity loss. In this context, there are basically two reasons to convert a preserved natural environment and incur the cost of reduced services from the environment, including the loss of biodiversity: natural environments are converted either to enhance production and consumption of material goods and services, or as an investment in an alternative asset that earns a greater return to the party responsible for the conversion.

Some conversion of natural environments is necessary to produce the material goods and services demanded by society. Production can be consumed or invested in other productive assets, either physical or human capital. On the consumption side, people need food, clothing, shelter, etc., and natural environments provide resource inputs for the production of material goods and services. Efficient resource allocation at a given point in time requires the marginal consumption value of the flow of productive inputs to equal the marginal value of preserved natural environments. The former depends upon the marginal value of consumption and the marginal productivity of the productive inputs. The marginal value of preserved natural environments is the discounted present value of the marginal amenity value, including biodiversity, plus any scarcity rent on the productive inputs provided by the natural environments (Krautkraemer, 1985; Barrett, 1992).

Intertemporal efficiency in the accumulation of assets requires equal

---

[2] This makes one cautious about long-term sustainability of harvesting non-timber products from tropical forest land. Ryan (1992) noted, 'removing industrial quantities of any material from an ecosystem is likely to cause major changes . . . neither group ("farmers" or government) has much experience in sustainable commercial use of tropical ecosystems.'

[3] The specific contributions of biodiversity are well documented in the literature. Wilson (1992) has provided a very readable compilation.

rates of return to each asset, so the rate of return to preserved natural environments should equal the rate of return to other assets.[4] The rate of return to an asset includes any capital gain and the asset's own rate of interest. The own rate of interest to natural environments is determined largely by the relative marginal amenity value, including the social return to biodiversity. Of course, efficiency requires that all values of preserved environments be taken into account by the decision-maker. As discussed below, failure to incorporate all values results in overly rapid depletion of preserved natural environments.

The growth-theoretic framework, then, helps to identify and categorize factors affecting conversion of natural environments – those factors that affect the marginal value of consumption, the marginal productivity of resource inputs, or the amenity value of natural environments. Important factors include economic growth, including development of alternative sources of consumption, population growth, and the static and dynamic incentives governing the use of natural resources. Economic growth can have both positive and negative effects on environmental preservation. The development and availability of alternative sources of consumption affect the marginal value of consumption and therefore the demand for productive inputs from natural environments. Environmental assets do not fare well in growth models where consumption goes to zero for technological reasons (e.g. the elasticity of capital-resource substitution or rate of technological progress is low). Indeed, the current threat to biodiversity seems greatest in those places where people are poor. If consumption can be maintained, then the amenity value of biodiversity can be large enough to preclude conversion of natural environments. This would include the case of an infinite marginal value attached to preserved natural environments in order to comply with a moral imperative that any reduction in biodiversity is wrong.

There are many ways by which consumption can be maintained without converting preserved natural environments. The development of backstop technologies that use renewable resources and maintaining or improving the flow of consumption services from already developed areas can reduce demand for the consumption services of preserved environments (Krautkraemer, 1986; Barrett, 1992). Technological progress in commodity production and capital-resource substitution also enhance the productive capacity of the economy which decreases the marginal value of consumption. However, these factors also increase the marginal productivity of

---

[4] Maximization of discounted present value requires the rate of return to each asset to equal the intertemporal rate of discount (which can be zero).

resource inputs and so the effect on preserving natural environments is ambiguous (Krautkraemer, 1985). Technological progress that enhances the production of commodities from already developed land does induce greater environmental preservation (Barrett, 1992).[5]

Thus, the level of economic development and the availability of other assets – physical and human capital – affect the ability to preserve natural environments. While increasing output per capita may contribute to biodiversity loss directly through habitat conversion in order to produce material goods and services, it is also associated with lower rates of population growth and greater willingness to pay for pollution control and environmental preservation generally. Nevertheless, economic development occurs with increased agricultural specialization, and biodiversity is lost as agriculture becomes more specialized (Norgaard, 1988).

Population growth increases the demand for consumption that can lead to greater development of preserved environments. Humans require relatively large quantities of food and other resources. Vitousek *et al.* (1986) estimated that humans use nearly 40% of potential terrestrial net primary product. This estimate uses a population of five billion people and a daily per capita consumption of 2500 calories. The human population directly uses 3.2% of potential terrestrial net primary product, coopted use increases it to 19% (30% on land) and lost productivity brings it to 25% (40% on land). While coopted use and lost productivity need not be proportional to size of population, these estimates suggest that doubling the population could be catastrophic for biodiversity unless resource use per capita declines dramatically.[6] Of course, this raises the question of what drives the growth of population. Certainly income and income security matter – population growth is lowest in developed countries.[7]

Since the intertemporal discount rate weights future values relative to current values, it also plays an important role in determining the rate of

---

[5] Southgate and Clark (1993) noted, 'farmers and ranchers encroach very little on natural habitat in countries where crop and livestock yields have improved. Conversely, where productivity trends have been flat, increasing demand for agricultural commodities, brought on by population or income growth or expanded exports have led inevitably to land-use conversion. In a sense, this analysis suggests that raising agricultural yields allows a country to "buy" the time needed to bring population growth under control.'

[6] Barrett (1994) pointed out that population growth also can increase the demand for preservation. Whether the optimal level of preservation increases or decreases depends upon the elasticities of marginal utility of consumption and amenities, the rate of population growth and the productivity of capital. It also depends upon whether or not all preservation values are taken into account by the relevant decision-makers.

[7] From 1970 to 1990 population grew at a 0.7% annual rate in developed countries versus 2.19% in developing countries. The annual rate of population growth over the period 1990–2000 is forecast to be 0.46% for developed countries and 2.03% for developing countries (WRI, 1992).

conversion of natural environments. A high discount rate can lead to complete depletion of preserved environments, with huge biodiversity losses, and collapse of the economy even if continued economic growth and permanent preservation are possible (Krautkraemer, 1985). However, the discount rate also can have indirect effects on environmental preservation. A lower discount rate increases the present value of biodiversity but also increases the present value of other assets and, at least in some cases, this can induce more rapid depletion of natural environments (Krautkraemer, 1988; Rowthorn and Brown, 1995). Gillis (1991) noted that a large share of deforestation has occurred as a result of government investment in infrastructure, electric power, and resettlement projects that low rates of discount make more appealing. As Krutilla (1967) noted, a lower discount rate 'may serve only to hasten the conversion of natural environments into low-yield investments'.

Since conversion of natural environments depends upon the relative value of alternative uses, an additional important factor is whether or not amenity values of natural environments are taken into account by the relevant decision-makers. The full preservation value will not be considered if biodiversity and other amenities are public goods whose benefits cannot be appropriated by individual decision-makers.[8] In this case, the rate of conversion of preserved natural environments and therefore the rate of biodiversity loss will be greater than is socially desirable (Krautkraemer, 1985).

The incentives facing decision-makers have a variety of static and dynamic effects, including an effect on the path of technological development. The high yields of modern agriculture result from crop varieties bred for yield rather than ecological stability.[9] Biodiversity is protected only to the extent that its benefits can be appropriated; otherwise, not all costs of specialized agriculture enter the many decisions that determine the development of agricultural technology. With no private return to biodiversity, we

---

[8] Sedjo (1992) noted 'having no unique claim to the returns of the genetic information embodied in the wild plants, individuals and countries engaged in developing the land resources will tend to ignore the potential economic value of the existing habitat as a repository for potentially valuable genetic resources. The destruction of genetic resources becomes an unintended consequence, an externality, of habitat-destroying land-use changes, and the costs of investing in protection and preservation can become substantial.'

[9] Lugo *et al.* (1993) noted, 'inverse relation between ecosystem complexity and net primary productivity explains why monocultures are necessary when the management objective is to maximize net yield and profit.' Flint (1992) noted, 'The main reason why diverse extractive systems and subsistence agricultural systems are replaced by more specialized agro-ecosystems is because they are inherently less productive. The important qualification to make is that even specialized agro-ecosystems benefit from local habitat diversity (wild pollinators, natural predators, etc.) and from management which maintains diversity, such as crop rotation.'

can expect technological development that is not mindful of biodiversity. If specialization allows increased output per unit of developed land, then, at least in principle, it is possible to leave more land in its natural state. Even here, the public good aspect of biodiversity plays a role. Incentives to increase output from already converted lands may be inadequate because the increased scarcity of biodiversity is not reflected in market prices.

The incentive structure affects even the selection of sites for environmental preservation. Scenic and recreational values may dominate ecological considerations, perhaps because individuals can benefit more directly from scenic and recreational preserves. In the United States, national parks tend to be selected for their scenic grandeur and recreational opportunity rather than for their biodiversity. Recreational groups object to the reintroduction of grizzly bears into the wilderness areas in the mountains of northern Idaho and eastern Washington. In Hawaii, the areas of greatest avian diversity are outside the protection of preserves (Winckler, 1992).

Biodiversity may suffer more than some public goods associated with preserved environments because it is a global public good. Protection is more difficult if the benefits of protection accrue globally and costs are borne locally (Wells, 1992; McNeely, 1993). Sedjo (1992) noted, 'it is much more difficult for the state to capture the returns to a global public good such as genetic resources . . . international law does not recognize property rights to wild species or wild genetic resource genotypes, and hence any rents . . . typically cannot be captured through domestic management . . .'. International cooperation is necessary to achieve global efficiency but such cooperation may not be self-enforceable (Barrett, 1995).

There also is an important intertemporal aspect to the public good aspect of biodiversity. The lack of incentive to provide public goods can directly affect the mix of assets individuals bequeath to future generations – in the absence of collective action, there is a bias to bequeathing private rather than public assets. An individual may recognize that a rich natural environment is a valuable asset for future generations but, since biodiversity and other global environmental assets are public goods, no individual can supply these goods to the next generation through individual actions. However, an individual can affect the amount of stocks, bonds, human capital, and other private assets given to his or her children and grandchildren.

Land tenure is another important aspect of the incentive structure. Property rights can be attenuated in time and/or in terms of rights to various products. There are numerous examples of the effect of attenuated property rights on environmental preservation. The 20-year lease on forest concessions in Indonesia provides insufficient incentive to replant or make other long-term investments since the lease period is shorter than the 35-year forest regeneration cycle (Barbier *et al.*, 1991). Peters *et al.* (1989) found that the net return to non-wood forest products in a tropical Peruvian region was greater than the timber value. However, the return to non-wood resources went to local, subsistence farmers while the government controlled timber export and earned substantial foreign exchange from timber sales.

Government policies also affect the incentives facing individual decision-makers. In principle, government intervention can correct market failures to reflect all costs of developing natural environments. However, in practice, governments also can fail to provide the socially desirable level of biodiversity protection. Indeed, in many cases, governments actually subsidize biodiversity loss. Governments do not operate in a vacuum but respond to political pressures generated by rent-seeking individuals and groups. The development of preserved natural environments and loss of diversity may have concentrated benefits and diffuse costs over both time and space.

There are numerous examples of government failure to protect biodiversity.[10] Cattle ranching has been responsible for 38–73% of deforestation in Brazil. The benefit:cost ratio for cattle ranching is 2.5 with government subsidies, but only 0.9 without subsidies (Barbier *et al.*, 1991). The value of logging in Indonesia is less than the value of non-wood forest products, and in the Bacuit Bay area of Palawan Island in the Philippines the cost of logging to fishing and tourism is four times the benefit of logging (Gillis, 1991). Governments in developed countries also fail to protect biodiversity even when non-environmental development costs exceed development benefits. During the US litigation concerning the Tellico dam, development continued not because of inattention to biodiversity but in spite of potential loss of a fish called the snail darter (*Percina tanasi*) and a negative net economic return (Davis, 1988).

The adverse impacts of government policies on biodiversity are not always direct. Trade policies can have important indirect effects on both the home country and its trading partners. Export subsidies for agricultural commodities in developed countries encourage overproduction in those countries – too much land cultivated intensively with excess use of

---

[10] Gillis (1991) noted, 'Nowhere in environmental policies has the role of prices in conservation been more neglected than in policies affecting the use of tropical forests. In nearly all countries, governments as owners have consistently sold wood resources at prices well below their commercial and social value.'

pesticides and fertilizers – while keeping world prices low and deterring agricultural investments in developing countries that are necessary for sustainable agricultural development.[11]

Relative magnitude of various factors

A growth accounting framework can be used to examine the relative magnitude of various factors affecting biodiversity loss. What follows is intended only as a rough guide to the relative magnitude of various factors. There are many crucial assumptions, both explicit and implicit, the data are not the best, and aggregation problems abound.

The number of species, S, in an area of size A, is often estimated by the equation:

$$S = CA^z, \tag{2.1}$$

where C and z are positive parameters with z ranging between 0.10 and 0.35 (Wilson, 1992). If $z = 0.25$, then the rate of change in the number of species is one-quarter of the rate of change in the size of the habitat.

Although extrapolation of this relationship between species and habitat size to species and preserved natural environments is not entirely correct, it can be used to give some idea of the magnitude of different factors affecting biodiversity loss.[12] Let D denote the amount of developed land, P denote the amount of preserved habitat, N denote population, and Q denote economic output (GDP). This gives the identity:

$$D \equiv (N)(Q/N)(Q/D)^{-1}. \tag{2.2}$$

Taking the logarithm of both sides and differentiating with respect to time (i.e. $dD/dt = \dot{D}$, etc.) gives:

$$\dot{D}/D = \dot{N}/N + \dot{q}/q - \dot{d}/d, \tag{2.3}$$

where q denotes economic output per capita and d denotes the ratio of economic output to developed land.

Equation (2.3) decomposes the rate of change in developed land into three factors: population growth, economic growth per capita, and the rate of change in output per unit of developed area. The rate of development of

---

[11] Wilson (1992) noted that government subsidies to developed-world farmers total $300 billion per year or six times the official foreign aid to Third World countries.

[12] The pattern of land development also matters. The rate of species loss is greater for the same habitat conversion if development is fragmented.

land is related to the rate of conversion of preserved environments by the following equation:

$$\dot{D}/D = -(\dot{P}/P)[P/(L-P)], \qquad (2.4)$$

where L denotes the total amount of land ($L = D + P$).

It is not possible to get firm figures on many of these factors and, in many cases, local rates of growth are more significant than global rates. Nevertheless, an examination of some rough estimates might be useful. The data used in the following examples are from the World Resources Institute (WRI, 1992).

World population growth is about 1.7% per year. Aggregation of economic growth rates across countries is difficult, but annual global economic growth per capita is certainly no greater than about 2%. The global rate of deforestation is about 0.3% per year. Using forest and woodland plus wilderness as the measure of preserved natural environments, we have a ratio of preserved to developed land, $P/(L-P)$, of 1.37. Combined with a deforestation rate of 0.3%, this gives a rate of increase in developed land of 0.41%. Consequently, it would seem that at the global level, the output per unit of developed land is increasing over time ($\dot{d}/d$ = 3.3%, if economic growth is 2%). It may be that market land values are increasing enough that land is being conserved as an input. This does not mean, however, that the rate of increase in output per unit of developed land is as great as it would be if all values of preserved land were considered.

The aggregate effects can conceal important differences across countries – for instance, Brazil is quite different. The Brazilian population grew at an annual rate of 2.07% over the period 1985–1990, and the GDP annual growth rate was 2.7% over the period 1979–1989, so per capita GDP grew about 0.7% per year. The annual rate of deforestation is approximately 0.4%–0.5%. Again using forest and woodland plus wilderness as the measure of preserved natural environments, $P/(L-P) = 8.61$ for Brazil. In Brazil, then, output per unit of developed land is decreasing ($\dot{d}/d$ = -1.5%).[13]

In the United States, annual population growth was 0.93% over the period 1981–1990 and annual per capita economic growth was 1.7%. The rate of deforestation was 1.1% and the estimated ratio of preserved to

---

[13] The story also can differ markedly across regions within a country. Myers, (1988) noted population in the southern sector of the Brazilian Amazon increased about 10-fold between 1975 and 1986, while land clearing increased from 1250 square kilometers to 17 000 square kilometers, more than 13 times.

developed land ($P/(L - P)$) is about 0.585. Consequently, output per unit of developed land is increasing in the USA at a good rate, probably greater than 3%. The figures for Japan give an increase in output per unit of developed land of 4.1%. It is tempting to suggest that developed countries are importing productive inputs from converted environments in developing countries.

Conclusion

The loss of biodiversity is driven by human conversion and degradation of natural habitat. Major factors affecting the rate of biodiversity loss are population growth, economic growth, and the public good aspect of many biodiversity benefits. Because biodiversity is a public good, not all benefits of biodiversity conservation accrue to those who bear the cost of conservation. Consequently, natural environments are converted more rapidly than is socially desirable. Changes in incentive structures are necessary for the full value of biodiversity to be taken into account by decision-makers, including governments. This could include taxes on land conversions, increased land preserves, and extension of property rights mechanisms. It also is important to provide alternatives to habitat conversion as a source of livelihood and to develop technologies that increase output per unit of developed land in order to reduce the pressures from population and economic growth to convert the remaining preserved environments that protect biodiversity. The difficulty of achieving needed changes in incentive structures should not be underestimated.

References

Barbier, E., Burgess, J. and Markandya A. (1991). The economics of tropical deforestation, *Ambio*, **20**, 55–8.

Barrett, S. (1992). Economic growth and environmental preservation. *Journal of Environmental Economics and Management*, **23**, 289–300.

Barrett, S. (1995). On biodiversity conservation. In C. Perrings, C. Folke, C. S. Holling, B. O. Jansson and K. G. Mäler, eds., *Biodiversity Loss: Theoretical and Ecological Issues*, pp. 283–97. Cambridge University Press, New York.

Davis, R. (1988). Lessons in politics and economics from the snail darter. In V. K. Smith, ed., *Environmental Resources and Applied Welfare Economics: Essays in Honor of John V. Krutilla*, pp. 211–36. Resources for the Future, Washington, DC.

Flint, M. (1992). Biological diversity and developing countries. In A. Markandya and J. Richardson, eds., *Environmental Economics: A Reader*, pp. 437–69. St Martin's Press, New York.

Gillis, M. (1991). Economics, ecology, and ethics: mending the broken circle for tropical forests. In F. Bormann and S. Kellert, eds., *Ecology, Economics, Ethics: The Broken Circle*, Yale University Press, New Haven, CN.

Krautkraemer, J. (1985). Optimal growth, resource amenities and the preservation of natural environments. *Review of Economic Studies*, **52**, 153–70.

Krautkraemer, J. (1986). Optimal depletion with resource amenities and a backstop technology. *Resources and Energy*, **8**, 133–49.

Krautkraemer, J. (1988). The rate of discount and the preservation of natural environments. *Natural Resource Modeling*, **2**, 421–37.

Krutilla, J. (1967). Conservation reconsidered. *American Economic Review*, **57**, 777–86.

Lugo, A., Parrotta, J. and Brown, S. (1993). Loss of species caused by tropical deforestation and their recovery through management, *Ambio*, **22**, 106–9.

McNeely, J. (1993). Economic incentives for conserving biodiversity: lessons for Africa, *Ambio*, **22**, 144–50.

Norgaard, R. (1988). The rise of the global exchange economy and the loss of biological diversity. In E. O. Wilson, ed., *Biodiversity*, pp. 206–11. National Academy Press, Washington, DC.

Peters, C., Gentry, A. and Mendelsohn, R. (1989). Valuation of an Amazonian rainforest, *Nature*, **339**, 655–6.

Rowthorn, B. and Brown, G. (1995). When a high discount rate encourages biodiversity. (Mimeo.)

Ryan, J. (1992). Conserving biological diversity. In *State of the World 1992, A Worldwatch Institute Report on Progress Toward a Sustainable Society*, pp. 9–26. W. W. Norton & Company, New York.

Sedjo, R. (1992). Property rights, genetic resources, and biotechnological change. *Journal of Law and Economics*, **35**, 199–211.

Solow, A., Polasky, S. and Broadus, J. (1993). On the measurement of biological diversity. *Journal of Environmental Economics and Management*, **24**, 60–8.

Southgate, D. and Clark, H. (1993). Can conservation projects save biodiversity in South America? *Ambio*, **22**, 163–6.

Vane-Wright, R., Humphries, C. and Williams, P. (1991). What to protect? – Systematics and the agony of choice. *Biology Conservation*, **55**, 235–54.

Vitousek, P., Ehrlich, P. R., Ehrlich, A. H. and Matson, P. A. (1986). Human appropriation of the products of photosynthesis. *BioScience*, **36**, 368–73.

Weitzman, M. (1992). On diversity. *Quarterly Journal of Economics*, **107**, 363–405.

Wells, M. (1992). Biodiversity conservation, affluence and poverty: mismatched costs and benefits and efforts to remedy them. *Ambio*, **21**, 237–43.

Wilson, E. O. (1992). *The Diversity of Life*. Harvard University Press, Cambridge, MA.

Winckler, S. (1992). Stopgap measures. *The Atlantic Monthly*, January.

WRI (World Resources Institute) (1992). *World Resources 1992–1993*. Oxford University Press, New York.

3

Biodiversity, economic growth and the discount rate

BOB ROWTHORN AND GARDNER BROWN

Introduction

Species extinction is a topic of grave concern to many who see it as a signal of our irresponsible stewardship of life on earth and perhaps even a harbinger of our own doom. Economists may be more cavalier about the issue for we know about optimal rates of capital accumulation, recognise that species are natural or genetic capital and that species with their own or intrinsic rates of growth below market rates of interest require a special pleading for their continued existence. Nevertheless, most of us would wonder about the legitimacy of an economy in which it was optimal to destroy species continuously and systematically. One natural setting in which to raise this question is a standard neoclassical growth model (Solow, 1976). We sketch the approach and results in the introduction in recognition of the fact that many readers will not have the patience to wade through the formal development in the main body of the text.

Individual utility is maximised as a result of an optimal investment programme over time, which produces optimal consumption in the light of a given technology subject to neutral technological progress and exogenous population growth. These are the usual assumptions and we are simply following tradition in order to focus attention on, and make a contribution to, elsewhere.

We make three innocuous innovations to the standard model. First, we acknowledge that human beings derive utility from the existence of other species. Individuals may enjoy visiting species *in situ*. In our model no distinction is made between the rareness of one species relative to another, so viewing value is not specific to any subset of species. Individuals may also find it fulfilling to know that species will be around now and in the future so

25

all may enjoy them – the more the merrier, according to our model. Thus, existence and bequest motives are the source of non-use values here.

Second, we harken back to Adam Smith and assume that a nation's output depends on land as well as capital and labour. Finally, we introduce an auxiliary 'production function' expressing the standard relationship between land availability and species richness. Species are more numerous when there is more land but there are diminishing returns (MacArthur and Wilson, 1967).

When the economy starts out there may not be enough machines, buildings and other forms of capital, compared to land and labour, so the economy rapidly builds up its capital stock, at the sacrifice of present consumption. Ultimately the economy reaches a steady state that has two important features. First, it is a dynamic steady state with per capita consumption generally growing, fuelled by a constant rate of technical progress. Output is growing enough to satisfy both a growing population and rising living standards. Second, and more interesting for biodiversity, the amount of land used in the production process eventually stabilises and the remaining land area is left undeveloped as a permanent reserve for species.

To summarise thus far, our results mimic the standard paradigm, with output and capital growing at a constant rate made up of the population growth rate plus the rate of neutral technical progress scaled by the way the production technology has been specified. Per capita growth rate in consumption is just the output growth rate reduced by the growth rate of the population. Land available for species preservation, to be enjoyed by all, stabilises at some optimal absolute level. The more we value species – are less willing to substitute produced goods for enjoying species – the more land is preserved. It is also not surprising that elements contributing positively to the underlying economic growth rate, such as neutral technical progress, are anathema to more biodiversity.

There are two further results, both counter-intuitive. First, there are cases when a high population growth rate leads to more land development, but there is one instance when this is not true. When output is produced under a constant return to scale technology in capital and labour (the doubling of both capital and labour doubles output), and the rate of interest is low, then more land is *preserved* under a faster population growth rate. The reason is that a larger population induces the substitution of labour for other inputs, including land, because labour becomes very much cheaper when it is more plentiful.

Hotelling simply confirmed the intuition of good micro-economists when he proved that higher discount (interest) rates imply greater rates of extraction of non-renewable resources and smaller equilibrium stocks of renewable resources.[1] The owner of natural resources can be expected to substitute out of them and into other forms of asset in response to an increased interest rate until the rates of return on all forms of asset are equalised. In the present case, this would suggest that a higher discount rate must automatically lead to the eventual extinction of a greater quantity of species. However, such a conclusion may not be correct. Hotelling's result arises from a partial equilibrium analysis in which the trajectory of the economy as a whole, and hence the demand functions for natural resources, are taken as given. As Pearce and Turner (1990) have pointed out, a higher discount rate will normally reduce the overall level, or growth rate, of output in the economy, thereby helping to restrain the demand for natural resources. Thus, the substitution and output effects of a higher discount rate pull in opposite directions, and the ultimate effect on natural resource use depends on the comparative strength of these effects.

For certain parameter values in our model, a higher discount rate will increase the cost of capital relative to land and encourage the substitution of land for capital. However, the higher discount rate also restrains the growth of output. The strength of this output effect is so great in our model that it *always* dominates any substitution effect, with the final result that less land is used in production and more species are preserved. It is interesting to note that the effect of a higher discount rate is to reduce the amount of both capital stock and land per worker employed. Thus, production becomes more labour intensive and this is ultimately the reason why both output and land use are depressed when the discount rate is increased.

To implement the optimal relationship between land use and the social discount rate presupposes either a wise planner or an institutional fabric that sees to it that those making the decision to develop the land are guided precisely by the public preference for species. Since these are non-use values, accurate signals are not and will not be forthcoming from straightforward markets. It will take more imaginative institutional design to get the optimal quantity of these (non-use) public goods.

The model

There are many measures of biological diversity. One of the most primitive is to regard species as a single homogeneous good. This approach is naive

---

[1] Throughout this chapter the term 'discount rate' refers to the discounting of future utilities. An increase in this rate implies a higher return on capital investment and a higher discount rate on goods.

for two reasons, one biological and the other economic. Only rarely are individuals indifferent between one species and another. Biologically, species differ in habitat requirements, in rareness and in uniqueness, to mention but three distinctions. Nevertheless, species enter the following neoclassical model in a primitive way.

We draw on the work of Robert MacArthur and E. O. Wilson who nearly 40 years ago joined cross-disciplinary intellectual forces to create a pathbreaking theory of island biogeography. Its elements include a theory of equilibrium of species that involves determinants of rates of immigration, extinction and successful colonisation strategies.

In the end, MacArthur and Wilson developed a stunningly simple – and familiar relationship; namely, that there is a log-linear relationship between the area of islands (I) and the number of species (S) expected to survive:

$$S = bI^{\gamma}, \tag{3.1}$$

where b and γ are constants. b depends on geographical characteristics and the taxon studied. In our model, this relationship is expressed in a slightly different, but mathematically equivalent, way. We normalize b to 1 and assume that S is given by:

$$S = h(A) \tag{3.2}$$

where:

$$h(A) = (1 - A)^{\gamma}, \quad \gamma > 0;$$

A being the proportion of total land area developed for use in agriculture and other forms of production. Thus, undeveloped land sustains species and functions like an 'island' in the MacArthur–Wilson sense.

Output (Y) in the economy requires capital (K), labour (L) and land (A) in time (t) and is blessed with Hicks' neutral technical progress:

$$Y = f(t, K, L, A). \tag{3.3}$$

We adopt the standard production function:

$$f(\cdot) = e^{\lambda t} K^{\alpha} L^{\eta} A^{\beta}, \quad 1 > \alpha + \beta, \text{ and } 1 > \alpha + \eta,$$

where λ is the constant rate of technical progress. In this economy, developed land is indispensable for output, so some species necessarily will have to be extinguished for a viable economy containing people. Output is allocated either to investment $\dot{K} = \dfrac{dK}{dt}$, where the 'dot' superscript refers to the time derivation of the variable in all cases, and consumption is denoted by CL, where C is per capita consumption. Thus:

$$Y = CL + \dot{K}. \tag{3.4}$$

Finally the utility function depends both on per capita consumption and species:

$$U(C,S) = C^{1-\sigma_c} S^{1-\sigma_s}, \quad \sigma_c + \sigma_s > 1, \quad 1 - \sigma_c > 0, \quad 1 - \sigma_s > 0. \tag{3.5}$$

Species are enjoyed for their non-consumption use value in this model. Non-economists can recognise that the marginal utility of species tends to infinity as the number remaining becomes smaller.

The manager of this neoclassical economy maximises intertemporal utility discounted at a constant rate, r, subject to the above conditions. Time subscripts are suppressed unless clarity requires specification. That is:

$$\underset{C,A}{\text{Max }} W = \int_0^\infty e^{-rt} \, U(C, h(A)) \mathrm{d}t, \tag{3.6}$$

subject to:

$$\dot{K} = Y - CL,$$
$$K(0) = K_0,$$
$$L(t) = L_0 e^{\ell t} \quad \ell \text{ exogenous,}$$
$$0 \le CL \le Y,$$
$$0 \le A \le 1.$$

The solution

It is shown in the appendix to this chapter that the solution to the above problem converges to a dynamic stationary state in which consumption, output, capital stock and population all grow at constant rates. Meanwhile, land use stabilises and a certain fraction is left permanently undeveloped. Let us define:

$$g^* = \frac{1}{1-\alpha} [\lambda - \ell(1 - \alpha - \eta)] \tag{3.7}$$

$$A^* = \frac{\mu(r + \sigma_c g^* + \ell)}{(1+\mu)(r + \sigma_c g^* + \ell) - \alpha(g^* + \ell)} \tag{3.8}$$

where:

$$\mu = \frac{\beta(1-\sigma_c)}{\gamma(1-\sigma_s)} > 0$$

$$k^* = \left(\frac{\alpha L_0^\eta A^{*\beta}}{r + \sigma_c g^* + \ell} \right)^{\frac{1}{1-\alpha}} \tag{3.9}$$

$$y^* = k^{*\alpha} L_0^\eta A^{*\beta} \tag{3.10}$$

$$c^* = \frac{\mu}{L_0} \frac{1-A^*}{A^*} y^* \tag{3.11}$$

Along the optimal path, the behaviour of the key variables is as follows:

$$A(t) \to A^* \tag{3.12}$$

$$\log K(t) \to \log k^* + (g^* + \ell)t \tag{3.13}$$

$$\log Y(t) \to \log y^* + (g^* + \ell)t \tag{3.14}$$

$$\log C(t) \to \log c^* + g^* t \tag{3.15}$$

By assumption:

$$\log L(t) = \log L_0 + \ell t \tag{3.16}$$

Thus, on the optimal path, the growth rates of K, Y and C eventually stabilise at $(g^* + \ell)$, $(g^* + \ell)$ and g^* respectively, whilst L grows at rate ℓ throughout. The behaviour of $\log K$ along the optimal path is shown in Figure 3.1. It is interesting to note that if $K_0 < k^*$ there is initially a very rapid growth in capital stock, which decelerates to the long run rate $g^* + \ell$. If $K_0 > k^*$ there is an initial period of slow growth.

Effect of parameter variations

It is straightforward to show that:

$$\frac{dg^*}{dr} = 0, \text{ and } \frac{dk^*}{dr}, \frac{dy^*}{dr}, \frac{dA^*}{dr} < 0. \tag{3.17}$$

Thus, the long-term growth rates of capital stock and output do not depend on the rate of discount, but their absolute levels do. The amount of land used for development also depends on the discount rate. Figures 3.2 and 3.3 illustrate the optimal paths of output and land development. The effect of a higher discount rate is to reduce the amount of capital stock and land used in production. Since the amount of labour employed at any time is

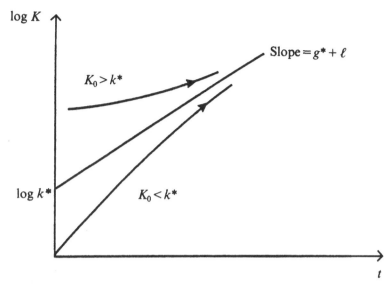

Figure 3.1. Optimal path for log K.

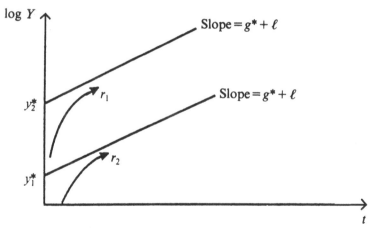

Figure 3.2. Optimal path for log Y ($r_2 > r_1$).

exogenous, the result is more labour-intensive production. In effect, the higher discount rate leads to the substitution of labour for both land *and* capital stock. Its effect on the capital:land ratio is ambiguous and depends on the precise parameter values.[2]

---

[2] $d(k^*/A^*)/dr > 0$ if $1 - \alpha - \beta$ is large; otherwise it is negative.

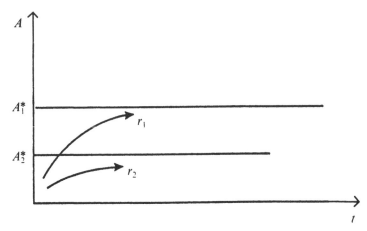

Figure 3.3. Optimal path for land development $(r_2 > r_1)$.

From equation (3.8) it follows immediately that $\mathrm{d}A^*/\mathrm{d}\lambda > 0$. Thus, unbiased technical progress leads to the eventual development of additional land. More interesting, perhaps, is the fact that under some conditions population growth may ultimately lead to less land use. This will occur, for example, if $1 - \alpha - \eta = 0$ and $r < (1 - \sigma_c)g^*$. In this case, it can be shown that:

$$\frac{\mathrm{d}A^*}{\mathrm{d}\ell} < 0, \frac{\mathrm{d}}{\mathrm{d}\ell}\left(\frac{K(t)}{L(t)}\right) < 0. \tag{3.18}$$

Thus, faster population growth results in relatively more labour-intensive production methods, which require less capital stock per worker and use less land in total. As might be expected, per capita output and consumption are reduced.

Conclusion

We used the introduction to summarize the principal results of the chapter. We use the conclusion to raise some questions and to make some general observations.

First, while it may be analytically convenient to assume an exogenous constant growth rate in the population, empirically it appears to be a function of per capita income and further may be susceptible to change through public expenditure on information and education.

Second, how would our conclusions change if we allowed for biased technical progress? In our model, technical progress is unbiased, and the faster it occurs the greater is the amount of land ultimately developed for use in production. However, the situation would be different with a strong bias towards land-saving, for then faster technical progress would lead ultimately to less land use.

Third, economists may consider tastes as given. In reality, tastes are formed. Conservation organizations spend substantial sums of money informing us about why we should care more about species preservation and reassuring converts that our tastes are 'right'. So the parameters in our utility function regarding the importance of species are really variables that can be influenced by expenditures on 'education'. In contrast to our model, where there is no conflict amongst homogeneous consumers, in fact some people like other species, and some people may be quite frightened by wild creatures and their habitat, which for them symbolizes mystery, the unknown and a threat to one's physical or psychological well-being (Nash, 1967). In due course, intertemporal models featuring biodiversity will acknowledge a distribution of tastes for it and portray the dynamics of investing in strategies that alter the intertemporal complexion of tastes and thereby the equilibrium number of species preserved in the long run.

Finally, we need to think creatively about the discount rate, particularly since it plays such a surprising role in this model. For what reasons is it a parameter or a variable that public agencies should use their resources to influence? Given the difficulty of accurately assessing how valuable species are, public policy makers may be confident that there should be a limit to species extinction that translates into an upper bound on habitat destruction. Ciriacy-Wantrup (1965) called this 'The safe minimum standard'. If the upper bound is incompatible with the parameters of this model – if the observed growth process is producing an extinction rate that does not make sense – some form of intervention is in order. It is within the spirit of this macroeconomic model to use monetary policy or other mechanisms to alter the discount rate in the desired direction. In a recent lecture, Solow (1992) wondered why any discount rate is used at all except as a 'technical convenience'. 'The usual scholarly excuse – which relies on the idea that there is a small fixed probability that civilisation will end during any little interval of time – sounds far fetched' (*ibid.*, p. 10). Our model provides a possible answer to Solow's query. In this model a failure to discount future utilities might eventually result in an unacceptable rate of habitat destruction. By making the discount rate positive, we can move output growth onto a lower trajectory and reduce the demand for land for use in production. Thus, one rationale for discounting could be to preserve the environment!

This raises a more general question about the nature of human preferences and their role in economic models such as ours. The traditional approach of intertemporal modelling is to take preferences as given and complete, and then seek an optimal time path in the light of initial conditions, technology and the natural environment. However, as Koopmans (1965, p. 226) has pointed out, this approach may assume an unrealistic degree of rationality, especially with regard to preferences. Outside a rather narrow range, preferences are often vague and inconsistent, and they are shaped and made coherent through practical experience or reflection. One form of reflection is the thought experiment, which explores the implications of certain choices and leads us to modify or flesh out our preferences by confronting us with new or unexpected situations. In Koopman's words:

> The underlying idea of this exploratory approach is that the problem of optimal growth is too complicated, or at least too unfamiliar, for one to feel comfortable in making an *entirely* a priori choice of an optimality criterion before one knows the implications of alternative choices. One may wish to choose between principles on the basis of the results of their application. In order to do so, one first needs to know what these results are. This is an economic question logically prior to the ethical or political choice of a criterion.

For example, our contemporaneous preferences for consumption goods and species may be well defined, but we may have no clear idea about what rate should be used for discounting the future. We may also wish to impose an upper bound on the ultimate extent of habitat destruction. By experimenting with a suitably calibrated optimal control model, we may find a discount rate that keeps habitat destruction within the required limits, whilst also satisfying the normal Hamiltonian conditions for optimization and respecting our contemporaneous preferences. In this way, such a model can help us to define our rate of time preference and thereby fill in some of the gaps in our preference structure.

Appendix: Derivation of the optimum path

The transformed problem

Here, we indicate how to solve the problem described in the main text of the chapter. This is done in two stages. First we transform some of the variables to derive a new problem. Having solved the new problem, we then transform these variables back to obtain a solution to the original problem. Let us define:

$$g^* = \frac{1}{1-\alpha}\left[\lambda - \ell(1-\alpha-\eta)\right], \tag{3.19}$$

$$k(t) = e^{-(g^*+\ell)t}K(t), \tag{3.20}$$

$$y(t) = e^{-(g^*+\ell)t}Y(t), \tag{3.21}$$

$$c(t) = e^{-g^*t}C(t). \tag{3.22}$$

The use of lower case for k, y and c denotes the fact that these are transformed variables. Equation (3.3) can now be rewritten as:

$$y = k^\alpha L_0^\eta A^\beta. \tag{3.23}$$

From the definition of U it follows that:

$$U(c, h(A)) = c^{1-\sigma_c}(1-A)^{\gamma(1-\sigma_s)}. \tag{3.24}$$

Hence, from equation (3.5):

$$\begin{aligned} U(c, h(A)) &= e^{-(1-\sigma_c)g^*t}C^{1-\sigma_c}(1-A)^{\gamma(1-\sigma_s)} \\ &= e^{-(-\sigma_c)g^*t}U(C, h(A)). \end{aligned} \tag{3.25}$$

The original problem can be restated as follows:

$$\operatorname*{Max}_{c,A} W = \int_0^\infty e^{-(r-(1-\sigma_c)g^*)t}U(c, h(A))\mathrm{d}t, \tag{3.26}$$

subject to:

$$\begin{aligned} &y = k^\alpha L_0^\eta A^\beta, \\ &k = -(g^*+\ell)k + y - cL_0, \\ &k(0) = K_0, \\ &c \le cL_0 \le y, \\ &0 \le A \le 1. \end{aligned}$$

The current-value Hamiltonian for the above problem, after introducing the adjoint variable m associated with the constraint on capital accumulation, is:

$$H = U(c, h(A)) + m\left[-(g^*+\ell)k + y - cL_0\right]. \tag{3.27}$$

The necessary conditions for an interior maximum of H are:

$$\frac{\partial H}{\partial c} = 0 = \frac{(1-\sigma_c)U}{c} - mL_0, \tag{3.28}$$

$$\frac{\partial H}{\partial A} = 0 = \frac{-\gamma(1-\sigma_s)U}{1-A} + \frac{m\beta y}{A}, \tag{3.29}$$

$$\dot{m} = [r - (1-\sigma_c)g^*]m - \frac{\partial H}{\partial k}. \tag{3.30}$$

Now (3.30) can be rewritten, after calculating $\frac{\partial H}{\partial k}$, as:

$$\dot{m} = \left[(r + \sigma_c g^* + \ell) - \frac{\alpha y}{k} \right] m. \tag{3.31}$$

From equations (3.28) and (3.29):

$$cL_0 = \frac{\beta(1-\sigma_c)}{\gamma(1-\sigma_s)} \left(\frac{1-A}{A} \right) y, \tag{3.32}$$

or:

$$cL_0 = \left[\frac{\mu(1-A)}{A} \right] y, \tag{3.33}$$

where:

$$\mu = \frac{\beta(1-\sigma_c)}{\gamma(1-\sigma_s)}. \tag{3.34}$$

From equations (3.26) and (3.33):

$$k = y \left(1 + \mu - \frac{\mu}{A} \right) - (g^* + \ell)k. \tag{3.35}$$

Note that, since $cL_0 \leq y$, it follows from equation (3.33) that:

$$A \geq \frac{\mu}{1+\mu}. \tag{3.36}$$

Stationary solution

There is a stationary solution $k(t) = k^*$, $y(t) = y^*$, $c(t) = c^*$, $A(t) = A^*$ and $m(t) = m^*$. This is determined as follows.

For $m^* \neq 0$, $k^* \neq 0$ it follows from equation (3.31) that the stationary output:capital ratio is:

$$\frac{y^*}{k^*} = \frac{r + \sigma_c g^* + \ell}{\alpha}. \tag{3.37}$$

To calculate the optimal level of land developed, begin with equation (3.35). When $k=0$:

$$A^* = \frac{\mu\left(\frac{y^*}{k^*}\right)}{(1+\mu)\left(\frac{y^*}{k^*}\right) - (g^*+\ell)}. \tag{3.38}$$

From equations (3.37) and (3.38):

$$A^* = \frac{\mu(r+\sigma_c g^*+\ell)}{(1+\mu)(r+\sigma_c g^*+\ell) - \alpha(g^*+\ell)}. \tag{3.39}$$

From equation (3.23):

$$k^* = \left(\frac{L_0^\eta A^{*\beta}}{\frac{y^*}{k^*}}\right)^{\frac{1}{1-\alpha}}, \tag{3.40}$$

where y^*/k^* and A^* are given by equations (3.37) and (3.39). The remaining variables are determined as follows:

$$y^* = k^{*\alpha} L_0^\eta A^{*\beta}. \tag{3.41}$$

From equation (3.33):

$$c^* = \frac{\mu}{L_0}\left(\frac{1-A^*}{A^*}\right) y^*. \tag{3.42}$$

From equations (3.28) and (3.42):

$$m^* = \frac{(1-\sigma_c)(1-A^*)^{\gamma(1-\sigma_c)}}{c^{*\sigma_c} L_0}. \tag{3.43}$$

Dynamics

The phase diagram for the pair (m,k), illustrated in Figure 3.4, has been developed elsewhere.[3] The curve $\dot{m}=0$ cuts the curve $\dot{k}=0$ only once and from below. The intersection occurs in this diagram on the downward sloping part of the $\dot{k}=0$ curve. However, depending on the parameter values, it could also occur on the upward part of the curve. This would not affect the dynamics. The initial value of k is equal to K_0. If $K_0<k^*$ the

---

[3] The derivation is available from the authors upon request.

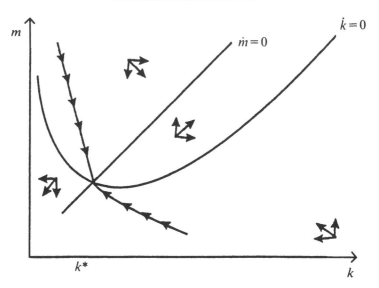

Figure 3.4. Phase diagram for m and k.

optimal path will involve a uniform increase in k towards its stationary equilibrium value. The opposite will be the case if $K_0 > k^*$.

The original problem revisited

Let $\hat{k}(t)$, $\hat{y}(t)$ and $\hat{c}(t)$ denote the optimal solution to the auxiliary problem. Using equations (3.20), (3.21) and (3.22) we can derive the following solution to the original problem:

$$K(t) = e^{(g^* + \theta)t}\hat{k}(t), \qquad (3.44)$$

$$Y(t) = e^{(g^* + \theta)t}\hat{y}(t), \qquad (3.45)$$

$$C(t) = e^{g^*t}\hat{c}(t), \qquad (3.46)$$

where g^* is given by equation (3.19).

The optimal path for $\hat{A}(t)$ is the same in each problem, since this is an untransformed variable. Note that $\hat{k}(t)$, $\hat{y}(t)$, $\hat{c}(t)$ and $A(t)$ converge to the limiting values k^*, y^*, c^* and A^* respectively. Thus, the solution to the original problem behaves in the fashion described in equations (3.8) to (3.11) of the main text.

References

Ciriacy-Wantrup, S. V. (1965). A safe minimum standard as an objective of conservation policy. In I. Burton and R. Kates, eds., *Readings in Resource Management and Conservation*, pp. 45–70. University of Chicago Press, Chicago.

Koopmans, T. (1965). On the concept of optimal economic growth. In *The Econometric Approach to Development Planning*, pp. 87–101. North-Holland, Amsterdam.

MacArthur, R. and Wilson, E. O. (1967). *The Theory of Island Biogeography*. Princeton University Press, Princeton, NJ.

Nash, R. (1967). *Wilderness and the American Mind*. Yale University Press, New Haven, CN.

Pearce, D. W. and Turner, R. K. (1990). *Economics of Natural Resources and the Environment*. Harvester Wheatsheaf, Hemel Hempstead.

Solow, R. (1976). *Growth Theory: An Exposition*. Oxford University Press, New York.

Solow, R. (1992). An almost practical step toward sustainability. Invited lecture on the occasion of the Fortieth Anniversary of Resources for the Future, Washington, DC, 8 October.

4

Uniformity in development and the decline of biological diversity

TIMOTHY M. SWANSON

Introduction

This chapter analyses the nature of the development process and its relationship to biodiversity decline. Here the focus is on 'development' in one particular sense: development as the rebalancing of the portfolio of societal assets (Solow, 1974). The specific issue is: what sort of developmental process would drive societies toward an increasingly narrow range of assets on which they rely? Hence, this is an analysis of biodiversity decline as the result of uniformity in development paths, both on a national and on a global basis. Why would a given society choose a narrow range of natural assets on which to base its economy, and why would other societies have incentives to emulate these choices?

The answer to these questions may lie within a recently initiated field of study termed 'endogenous growth theory' (Romer, 1987). The motivating concept here is that there are increasing returns to scale to certain forms of investments. The proponents of the theory argue that it is important to invest socially in those areas where increasing returns may justify those investments once a sufficient scale is attained. The theory has been used to prescribe investments in information-generating areas of the economy (such as certain services and high-technology sectors) because these are the areas replete with positive externalities.

The fundamental insight motivating this chapter is that it is equally possible to turn this theory to positive (as opposed to prescriptive) purposes, and to argue that the pattern of investment that has been observed in the past is explained in part by virtue of various forms of returns to scale. Then it is possible that the uniformity that is observed within development processes has been generated by the enhanced growth prospects to be derived from following previously initiated development paths, on account of the positive externalities conferred by the already-developed countries. This is what endogenous growth theory would imply: there is an in-built inertia toward uniformity in making choices regarding paths to development.

This chapter makes this argument within the context of a case study of a part of the biodiversity problem that is largely neglected throughout the remainder of the volume, i.e. the homogenisation of the varieties utilised in modern agriculture. This is a good case study because of the obvious importance of agriculture to human sustenance, and because of the clear problems associated with reducing the range of asset forms relied upon within agriculture. For example, human society now relies upon a mere four forms of plants (maize, wheat, potatos, rice) for the majority of its daily sustenance, and increasingly these four species of plants are being utilised in the form of a mere handful of specific high-yielding varieties. This represents a very narrow food pyramid on which to build an ever more substantial human society (Swanson *et al.*, 1994). What forces are causing human society to invest in so narrow a range of biological assets?

The points being made here may be generalised to the level of all of human development, not just agricultural development. The forces that drive agricultural development toward an increasingly narrow range of assets exist within all spheres of human activity. The object of this chapter is to demonstrate how the forces for uniformity within development paths drive the decline of biological diversity.

The chapter commences with a section that describes the link between development and land use conversion; it explains why developing societies rearrange the resources that were naturally endowed. The following section then explains how the land use practices known as modern agriculture are a particular type of conversion, where human societies have partnered very specific species in the process of development. The next two sections then reach the nub of the issue. They describe the externalities involved in the conversion decision that create incentives for the use of a small range of assets in human land use, and also create incentives for subsequent societies to make the same selections as the first. It is these externalities that generate the force for uniformity in the paths of development that are taken by each society. The forces for uniformity within human development generate the problem of diversity decline.

What is conversion?

One of the fundamental economic problems is the determination of the form of the assets in which human societies will hold their aggregate wealth (their 'portfolio'). If the optimal portfolio determined by this process is different from the initial, then there will be incentives to convert some existing assets to other forms. For societies with some endowment of natural resources, this may imply disinvestment in some naturally occurring assets, and investment in some other human-preferred ones.

The incentive to disinvest in the naturally occurring stock of assets in order to invest in the human-selected ones is referred to as the conversion process. Conversions occur because the natural form of any resource is necessarily in competition with all other forms in which humans might wish to hold the value of these same assets. The conversion decision derives from the fact that human societies now exercise the power to choose whether to hold natural resources in their original form, or to substitute another (Solow, 1974).

This conversion decision extends even to biological resources (Swanson, 1990). For example, a given hectare of land, which is originally growing diverse native grasses, may be converted directly to another grassy plant form such as wheat, because of the enhanced productivity of this resource. Alternatively (and less directly), a tropical forest may be logged and sold with the funds then invested in other societal assets (such as education), resulting in the conversion of the natural asset to, for example, human capital. Species must also be viewed as societal assets, because humans have demonstrated the capability to determine the continuing existence of most species on earth, either directly or indirectly through their land use decision-making. Thus, the continued existence of most species will depend upon their capacity to find a role in the societal asset portfolio.

This is the fundamental nature of the forces driving diversity decline: societal development by means of portfolio conversions. Diversity decline is caused by human reconstruction of the global portfolio of biological resources, from which most biological resources (the so-called diverse resources) are being excluded. Whenever a given species is not expressly selected for inclusion within this portfolio, it is subjected to the general forces for disinvestment that lead to its decline. These assets will be converted if they are inferior assets when compared with competing means of production (i.e. other species) in combination with the ancillary resources required for production (land and management). Thus, a species will implicitly be assessed in regard to its productive use of a given allotment of land or labour, and if the return from its use of these factors is not competitive then the resources will be allotted to other species' uses. The decline of most species is determined implicitly by the allotment of most human-controlled resources (land and management) to other species.

The link between land use conversions and diversity decline may be seen in the studies of the biodiversity problem: they demonstrate the link between the general problem of biodiversity decline and the 'conversion' of natural habitats. Estimates of aggregate natural habitat conversions over the past two centuries range from 25% to 50% of the original land area (IIED and WRI, 1989). Two hundred million hectares of forestland and 11 million hectares of natural grasslands were converted to specialised agriculture between 1960 and 1980 alone, all of it in the developing countries (Holdgate *et al.*, 1982). All of these data demonstrate that these societies are in the process of rebalancing their portfolios, away from the naturally endowed portfolio toward one selected by those societies. In the process, the natural portfolio is being discarded.

Land use conversion is directly linked by natural scientists to species extinction; the analysis of the relationship has itself spawned a field known as 'island biogeography'. Empirical studies have described a general relationship whereby the conversion of 90% of the area of a given natural habitat will result in the loss of half of its naturally resident species (MacArthur and Wilson, 1967). In addition, the popular perception of the biodiversity problem is itself derived from this link. All of the projections of 'mass extinctions' are based on extrapolations of land use conversion trends into the final refugia of species diversity, i.e. the tropical rainforests. A fairly conservative estimate would seem to be that the diffusion of these technologies into these final refugia are causing current rates of extinction to be about 1000 to 10 000 times the historical rate of extinction (Wilson, 1988).

These studies leave little doubt about the relationship between diversity and development. Development and conversion go hand in hand, and conversion is the process by which natural habitat and its resident species are lost. In the 'developed countries' of Europe and in the USA, there is little land conversion still taking place. The entirety of the land conversions for purposes of agricultural development between 1960 and 1980 (Table 4.1) occurred within the 'developing world'. This is because the process of land use conversion is largely completed in those parts of the world known as 'developed'. The proportion of Europe which is 'unmodified habitat' (of at least 4000 square kilometres in area) is now certifiably zero. In the USA, the proportion of unmodified habitat of this dimension is down to about 5% of the American land mass. This is to be contrasted with a global average of about 30% (WRI and IIED, 1990). This asymmetry in the holdings of natural habitat is the result of unevenly applied development. The land conversion process has worked its way across most of the

Table 4.1. *Rates of conversion of natural habitat to agriculture*

	1960 (million ha)	1980 (million ha)	% change
Developing			
Sub-Saharan Africa	161	222	37.8
Latin America	104	142	36.5
South Asia	153	210	37.2
S. E. Asia	40	55	37.5
Developed			
North America	205	203	0.1
Europe	151	137	0.0
USSR	225	233	0

Source: Repetto and Gillis (1988).

Table 4.2. *Recent rates of conversion to specialised agriculture (ten year rate – to 1987) (increase in land area dedicated to specified use)*

Conversions to cropland (%)		Conversions to pastureland (%)	
Paraguay	71.2	Ecuador	61.5
Niger	30	Costa Rica	34.1
Mongolia	31.9	Thailand	31
Brazil	27	Phillipines	26.2
Côte d'Ivoire	24	Paraguay	26.0
Uganda	21.4	Vietnam	14.0
Guyana	21.3	Nicaragua	11.8
Burkina Faso	19.4		
Rwanda	18.6		
Thailand	17.1		

Source: WRI and IIED (1990).

developed world, and it is now proceeding in similar manner in those countries now known as 'developing'. At the frontier of this technological diffusion, the rate of conversions remains high (Table 4.2).

Therefore, the diversity of life forms is in decline because large-scale changes in land use have been occurring across the face of the earth for many centuries, initially in the parts now known as the developed world, and currently in the developing. It is straightforward to identify the driving force behind these large-scale changes: the human conversion of the natural endowment of biological resources to a human-selected set of assets. It is clearly human capabilities and human choices driving biodiversity decline, in the pursuit of human development. The question remains: why is development being pursued in a manner that drives diversity into decline?

Conversion and 'agriculture'

The key to this explanation probably lies in a technological change that occurred originally about 10 000 years ago; this was the realisation of agriculture by human societies. Agriculture has consisted of the selection of a few prey species, and the expansion of their ranges. Prior to the occurrence of this idea, human societies preyed upon species over their natural ranges (hunting and gathering); afterwards, human societies transported the species they used, displacing the naturally selected varieties. The discovery of this strategy (domestication and cultivation) and its implementation constituted a very important part of a technological shift that occurred in the late Pleistocene. This was a process that was important to the advancement and development of human society and civilisation as we know it, but it is also a process that has generated the potential for a decline in biodiversity as a by-product.

Human advancement through agriculture has not been built directly upon diversity decline, in the sense of the overuse and/or mining of biomass. Rather, human advancement has come hand-in-hand with reliance upon a small set of species and the expansion of their ranges (at the expense of other species). The expansion of their ranges (with the simultaneous constriction of other prey species' ranges), and the consequent expansion of the human niche (with the simultaneous constriction of other predator species ranges), has resulted in the global homogenisation of the biosphere. It is this homogenisation that, on the one hand, has generated human development and, on the other, has generated the decline of diversity.

The earliest archaeological evidence of agriculture dates back to a period between 6000 and 10 000 years ago. This consists of the first signs in the fossil records that human societies were selecting individual species and translocating them with their culture. It is now the case that the biological production 'menu' for the bulk of all human society has converged upon a relative handful of species. Of the thousands of species of plants that are deemed edible and adequate substitutes for human consumption, there are now only 20 species producing the vast majority of the world's food. In fact, the four big carbohydrate crops (wheat, rice, maize and potatoes) feed more than the next 26 crops combined (Witt, 1985). Although it is estimated that humans have utilised 100 000 edible plants species over their history, barely more than 150 species are now under cultivation (Esquinas Alcazar, 1993). In short, humans have come to rely upon a minute proportion of the world's species for their sustenance; these species are termed here the 'specialised species'.

Human choice resulted in biosphere homogenisation not only at the lower trophic levels, by reason of human selection of prey species. It also resulted in homogenisation at higher levels, by reason of human population expansion (and by reason of the elimination of other predators' prey species). Thus, it was at this same time (about 10 000 years ago) that the population of the human species began to record unprecedented growth. The development of human technologies (cultivation and domestication) in the Neolithic period enormously expanded the human niche, from the capacity to support perhaps 10 million individuals to a capacity of hundreds of millions in a relatively short time (Boulding, 1981). Most palaeoarchaeologists date a substantial increase in human populations to this period (Biraben, 1979).

What then is agriculture? From an evolutionary perspective, agriculture is the human usurpation of the function previously performed by 'relative fitness'. Over the course of millions of years the allocation of a portion of life-sustaining base resources (e.g. land) has been determined in a natural competition between life forms, and the naturally endowed diversity was the observed outcome of this competition. In the past 10 000 years, however, this allocation decision has been usurped by the human species, as humans have become capable of determining the set of biological resources with which they live (rather than simply hunting or gathering those that existed naturally). Base resources are now allocated by humans to the various species that continue in existence. Hence, the constitution of the biosphere has become one part of the human developmental process. Agriculture is the process by which humans practice development with respect to the biosphere, but why do they practice it in such an exclusive manner? That is to say, what is the benefit to be gained by human societies from partnering so small a set of prey species to the neglect of all others?

Agriculture and homogeneity: the benefits to uniformity

Conversion as an economic force explains only why it is the case that the natural set of biological resources might be replaced by another human-selected set on any given parcel of land, depending upon relative productivities, but it does not explain why a small number of species would replace millions across the whole of the earth. In other words, this force implies conversion but not necessarily homogenisation. In order to explain the global losses of biodiversity, i.e. *a narrowing of the global portfolio*, it is necessary to identify the nature of the force that would generate this homogenisation of the global biosphere.

Specifically, it is unlikely that a wholly natural process would drive the world toward less diversity. This would require the evolution of both biological generalists (species with superior productivity across many niches) and uniform human tastes (across the globe). In fact, the current drive toward uniformity is contrary to the very idea of evolutionary fitness. Fitness implies competitive adaptation to the specific contours of a certain niche. The evolutionary process generates species that are well-adapted to their own specific niches through a process of niche refinement; that is, a surviving species represents a 'good fit' to its own niche (Eltringham, 1984).

It is equally unlikely that human tastes are so uniform as to demand the homogenisation of biological resources. Communities 'co-evolve' in order to fit better within the system in which they participate. It would be expected that the preferences of predators would be determined generally by their available prey species. In fact, there is ample evidence to support this expectation that human communities would prefer to consume the resources they depended upon traditionally (Swanson, 1990; Cooper, 1993).

This confirms that the depletion of diversity is not a natural phenomenon; rather, it is a socioeconomic one. The process of the selection of assets for society's portfolio is an economic decision, determined by forces that shape the perceived relative advantageousness of different assets. There is no reason to expect that nature would have evolved these biological 'generalists' that are now monopolising global production, because competitive adaptation and co-evolution militate against that conclusion. If it is not the naturally given characteristics of the domesticated and cultivated species that is determining their universality, then it must be some characteristic related to the economic system of which they are part. In this sense, the process of homogenisation of the biosphere is a wholly economic process, and not a natural one.

What is the nature of this process? It is being generated by the appropriation of the benefits accruing to uniformity. It is not just the particular characteristics of the species that are determining its selection, but also the uniformity of the variety itself. This is indicated by the fact that

Table 4.3. *The decline in diversity at the level of varieties*

Crop	Country	Number of varieties
Rice	Sri Lanka	From 2000 varieties in 1959 to 5 major varieties today
Rice	India	From 30000 varieties to 75% of production from less than 10 varieties
Rice	Bangladesh	62% of varieties descended from one maternal parent
Rice	Indonesia	74% of varieties descended from one maternal parent
Wheat	USA	50% of crop in 9 varieties
Potato	USA	75% of crop in 4 varieties
Cotton	USA	50% of crop in 3 varieties
Soybeans	USA	50% of crop in 6 varieties

Source: World Conservation Monitoring Centre (1992).

diversity is declining within species as much as it is across species. Human choice is focusing down to the most particularist level that plant and animal breeding operations allow (see Table 4.3).

Why is uniformity economically important? It is important because it is the characteristic of uniformity in the biological asset that renders it combinable with other factors of production in modern agriculture. That is to say, it is the common and stable characteristics of the modern variety that make it possible to design with particularity the capital equipment (planting and harvesting machinery), chemicals (fertilisers, pesticides), and all of the other ancillary requirements of modern mass-produced agriculture. If this were not the case, then the tools of agriculture would be required to conform to fields full of plants of varying sizes and shapes. Although mixed production of this nature might be optimal biologically (in order to reduce pest invasions, for example), the heterogeneity implied in the method would disallow the application of standard tools and technologies. It is the increasing returns from the use of methods of mass production in agriculture that drives the use of uniform varieties in production.

Therefore, the degree of homogenisation witnessed in modern agriculture is explicable by reference to the requirements of 'mass production'. Industries based on these techniques require that their inputs be standardised, so that they are readily combined, in a carefully controlled environment. Of course, this is precisely the nature of modern land conversion trends to agriculture. Land clearances occur in order to create a uniform base upon which to build the new production system. Then, the new system is created around a standard 'production method', constituting the particular variety and the tools that correspond to it. The entire monocultural production system is substituted for the diversity that existed previously, and this substitution generates the net loss in diversity.

The nature of modern agriculture generates the degree of homogenisation witnessed in modern farming systems because it is a mass-production-based system. Mass production requires uniform inputs and generates homogeneous outputs. This change in technique of production explains the decline in diversity seen within human agricultural systems.

The diffusion of agriculture

Even though mass production in agricultural methods requires standardised varieties for implementation, it is still necessary to explain why the same varieties of the same species are chosen for introduction in all parts of the globe. Why is agriculture adopted 'in whole' rather than adapted 'in concept' in many different environments across the face of the globe?

Homogenisation of the global biosphere requires some sort of non-convexity in the social choice set that would cause previous selections to determine later ones. That is to say, there is the necessity of a spillover between earlier decisions and later choices. The technological change known as agriculture was of this nature. The idea that originated about 10 000 years ago was centred on the idea of creating species-specific tools and technologies, and translocating a particular species in combination with its technology as a single 'method of production'. In other words, the idea concentrated on the development of the technologies for efficient agricultural production that were focused upon a single set of species. The result was the development of two new important factors of production in the production of biological goods: species-specific capital goods and species-specific learning. It is the combination of these factors, together with the specialised species, that generates the force for biosphere homogenisation. As a single method of production (species–tools–experience), the originally selected species are able to outcompete the naturally existing ones at any given location on account of the tools and experience that come with them. Agriculture came embedded in certain early-selected species (Swanson, 1994).

The accumulation of the capital goods specialised to these species goes hand in hand with the adoption of these species. This may be seen at the 'conversion frontier'. For example, over the past ten years, the number of tractors in Africa increased by 29%; they increased by 82% in South

America; and by 128% in Asia. During the same period the number of tractors decreased by 4% in North America (WRI and IIED, 1990). Societies that are introducing the specialised species do so in part because these species are tailored to the tools that are used with them. It is the combination of species and species-specific tools that constitutes a 'method of production'. When a conversion decision is being made, a country will consider all of the possible methods of production (species/capital goods combinations) in a search for the most efficient method.

The other important factor introduced into the production of biological goods was species-specific learning. With more experience with a particular species, it was possible to become even more efficient in its production (by reason of increased understanding of its biological nature, as well as intervention to determine the same). This information became another crucial factor for agricultural production, but it existed only in one form – embedded in the received forms of the domesticated and cultivated species.

It is the nature of this final factor that generates the forces for the convergence of the biosphere upon a small set of specialised species. It is the dynamic externality inherent within accumulated knowledge and learning that generates the non-convexity within the system, so that human choice falls again and again upon the same small set of life forms. Specifically, accumulated knowledge in this context is a *non-rival good* in the sense of Romer (1987, 1990*a*,*b*), i.e. it is of the nature of a 'design or list of instructions' that is distinct from the medium on which it is stored and thus (as pure information) it may be used simultaneously by many agents without added cost. The accumulated experience in regard to the specialised species is inherent within the capital goods and species as they stand, and is available at no added costliness to the marginal user (Romer, 1990*b*).

Agriculture originated in the Near East. It consisted of a set of ideas, a set of tools, and a set of selected species. At that time and in that locale, each of these selections was locally optimal. However, the set of ideas–technology–species was transported out of that region as a single unit, as the continuing investments in this combination caused the ideas and tools to become embedded in the chosen species.

A non-convexity was introduced within the decision-making process, by reason of the non-rival nature of the information embedded in the specialised species (that would be costly to produce for any diverse species). This is the essential difference between the specialised (domesticated) species and the diverse (wildlife) species. For one group, an information set is publicly available as an input into their production; for the other, it is necessary to construct that same information.

The global conversion process has consisted of the extension of these chosen species' ranges. As a consequence, much of the face of the earth has been reshaped in order to suit these few species and the tools used in their production. It is the diffusion of this 'bundle' of ideas–tools–species that is at the base of the biodiversity problem.

Just as the initial technological changes within agriculture narrowed the range of species used in biological production, the more recent changes have been narrowing the range of varieties of a species being used. Uniformity becomes increasingly important, as techniques and capital goods become increasingly focused. The agrichemical revolution of the past half-century has focused agriculture on a small number of varieties for which the characteristics of their predators and nutrient requirements are well known. The diffusion of modern agriculture became embedded within a method of production that consisted of the modern specialised variety, its chemicals and its capital goods (tools).

Conclusion

Biodiversity decline has been portrayed here as the outcome of uniformity within the development process when applied to the biosphere. Human societies realised the possibility of developing the biosphere, with the advent of agriculture. Since that time societies have chosen the portfolio of living assets on which they will rely, rather than using that which nature had allocated to that territory. The chosen species have become a part of the overall 'method of production' humans use in biomass production. As this same development strategy has diffused across the earth, it has resulted in the homogenisation of the biosphere, and the decline of diversity.

With the advent of modern agricultural industry, the forces that had shaped the selection of the small set of species that constituted the domesticated or cultivated species were applied at the level of intraspecies varieties. The larger investments in species-specific tools and technology required even greater uniformity in the biological asset used in the process. The green revolution has been driven by the diffusion of these mutually specialised methods of production: modern variety–specialised chemicals–specialised tools. The conversion of agricultural lands to the technologies implicit within the green revolution has meant that the world's agriculture has concentrated increasingly upon the same set of modern varieties.

Hence development has been in conflict with diversity over many centuries – but only by reason of the specific manner in which it has been practised. Uniformity in development exists because there are certain forms of knowledge accumulated with human experience that are readily communicated to the next developing state. This knowledge confers increasing returns to scale to uniform development because it is no more costly for all countries to use than it is for one. However, these returns to scale are only available if the technology is adopted in whole, as the accumulated experience is species specific. In other words, if all species reacted similarly to the chemicals, tools and methods employed by humans, then the experience would apply equally and there would be no necessary conflict between human development and biological diversity. It is the presence of meaningful diversity within the biosphere that results in the non-transferability of the accumulated experience across species boundaries. Thus, development remains biased towards the use of the same species, and the accumulated experience that goes with them. This is the force that drives uniformity in development choices, and correlatively the decline of diversity.

References

Biraben, J.-N. (1979). Essai sur l'évolution du nombre des hommes. *Population Bulletin*, 1.

Boulding, K. (1981). *Ecodynamics*. Sage, London.

Cooper, D. (1993). Plant genetic diversity and small farmers: issues and options for IFAD. *Staff Working Paper* no. 13. International Fund for Agricultural Development (IFAD).

Eltringham, S. K. (1984). Wildlife Resources and Economic Development. John Wiley & Sons, New York.

Esquinas-Alcazar, J. (1993). The global system on plant genetic resources. *Review of European Community and International Environmental Law*, 2, 151–7.

Holdgate, M., Kassas, M. and White, G. (eds.) (1982). *The World Environment 1972–1982*. United Nations Environment Programme: Nairobi.

IIED and WRI (International Institute for Environment and Development and World Resources Institute) (1989). *World Resources 1988–89*. Basic Books: New York.

MacArthur, R. and Wilson, E. O. (1987). *The Theory of Island Biogeography*. Princeton University Press, Princeton, NJ.

Repetto, R. and Gillis, M. (eds.) (1988). *Public Policies and the Misuse of Forest Resources*. Cambridge University Press, Cambridge.

Romer, P. (1987). Growth based on increasing returns due to specialisation. *American Economic Review*, Papers and Proceedings, 77, 56–62.

Romer, P. (1990a). Endogenous technological change. *Journal of Political Economy*, 98, 245–75.

Romer, P. (1990b). Are nonconvexities important for understanding growth. *American Economic Review*, Papers and Proceedings, 80, 97–103.

Solow, R. (1974). The economics of resources or the resources of economics. *American Economic Review*, 64, 1–12.

Swanson, T. (1991). Conserving biological diversity. In Pearce, D. (ed.), *Blueprint 2: Greening the World Economy*, pp. 181–208. Earthscan: London.

Swanson, T. (1994). *The International Regulation of Extinction*. Macmillan, London, and New York University Press, New York.

Swanson, T., Pearce, D. and Cervigni, R. (1994). *The Appropriation of the Values of Plant Genetic Resources for Agriculture*. Commission on Plant Genetic Resources, Food and Agricultural Organization, Rome.

Wilson, E. O. (1988). *Biodiversity*. National Academy of Sciences, Washington, DC.

Witt, S. (1985). *Biotechnology and Genetic Diversity*. California Agricultural Lands Project: San Francisco.

World Conservation Monitoring Centre (1992). *Global Biodiversity*. Chapman & Hall: London.

WRI and IIED (World Resources Institute and International Institute for Environment and Development) (1990). *World Resources 1990–1991*, Oxford University Press, Oxford.

Part B

Diversity decline as institutional failure

5

Decline in biodiversity and risk-adjusted net national product

JOHN M. HARTWICK

Introduction

A natural way to view biodiversity is as a homogeneous stock S with respect to time (t). Loss of biodiversity is then simply a shrinking of the stock. Matters can be made simple if the stock is identified with forested land. Then deforestation becomes the instrument causing a reduction in biodiversity. On the benefit side, matters assume a less crisp form. In one view the stock $S(t)$ is associated with a flow of positive services to consumers and/or producers. Less biodiversity is then like more pollution. Such a view leads us to identify biodiversity as an 'indirect' benefit[1] of having a forest and to propose taxing deforestation (or subsidizing forest owners) in order to internalize the unmarketed benefits provided by the biodiversity stock, $S(t)$. In the first part of this chapter, I formalize this line of thinking.

A different view of the benefits of biodiversity is that they are made manifest only when the biodiversity is gone. Society wakes up one day with an ecological 'crunch' on its hands. For example, agricultural activity might fall drastically because a key organism has gone extinct. Such structural breaks have been linked to pollution (e.g. Cropper, 1976; Heal, 1984). The view is that economic activity takes place normally but in fact is creating a trigger that one day shows itself in ecological collapse. The date of collapse or of structural break is a random variable changing over time. For example, hydrocarbon combustion can lead to a build-up of carbon dioxide and the cumulative emissions increase the probability of a structural break occurring. This view also leads to the need for a tax on emitting activity but in this case the tax is acting to delay the structural break and the tax is not functioning directly to yield cleaner air for current 'users'. With

---

[1] Pearce (1992) among others used the term 'indirect benefit' of a forest and links biodiversity to a group of indirect benefits.

biodiversity, we generally appreciate it only when we are forced to confront its loss. This view is taken up in the second part below.

Model I: Pigovian price assigned to land in forest

In this case, biodiversity gradually diminishes as deforestation takes place and the effect is a gradual decline in productive efficiency in the economy. One might view this as a gradual loss in soil productivity as biodiversity diminishes. The economy's production function is $\phi(S)Q(K, N, L)$ where $\phi(S)$ is an efficiency index and S is the amount of land in forest. S is also standing for the stock of diverse 'species'. L is land in agriculture; $L + S = S_0$, S_0 being total land. K is machine capital[2] and N is population size, assumed constant. Then:

$$\dot{K} = \phi(S)Q(K, N, L) - C - g(R), \tag{5.1}$$

where $\dot{K}$ etc. is dK/dt etc., C is aggregate consumption and $g(R)$ is the cost of shifting R hectares from forest to agriculture. Then:

$$R = -\dot{S}. \tag{5.2}$$

There will be an initial $K(0) = K_0, S(0) = S_0$ and $L(0) = L_0$. The savings/ consumption trade-off will emerge from the maximization of the present value of the utility of consumption in $\int_0^\infty U(C)e^{-\rho t}dt$. The current-value Hamiltonian for this problem is:

$$H(t) = U(C) + \lambda(t)(\phi(S)Q(K, N, L) - g(R) - C) - \mu(t)R, \tag{5.3}$$

where $l(t) = S_0 - S(t)$. Now:

$$\frac{\partial H}{\partial C} = 0 \rightarrow U_C = \lambda, \tag{5.4}$$

and:

$$\frac{\dot{\lambda}}{\lambda} = \frac{\dot{U}_C}{U_C}, \tag{5.5}$$

$$\frac{\partial H}{\partial R} = 0 \rightarrow \mu = -\lambda g_R, \tag{5.6}$$

$$-\frac{\partial H}{\partial K} = \dot{\lambda} - \rho\lambda \rightarrow \frac{\dot{\lambda}}{\lambda} = \rho - \phi(S)Q_K, \tag{5.7}$$

---

[2] Some leap is needed to imagine machine capital being generated in an economy specialized in agriculture but I will not linger over this non-essential detail.

$$-\frac{\partial H}{\partial S} = \dot{\mu} - \rho\mu \rightarrow \frac{\dot{\mu}}{\mu} = \rho - \frac{\lambda}{\mu}(Q\phi_S - \phi Q_L). \tag{5.8}$$

We also require:

$$\lim_{T\to\infty} \lambda(T)K(T)\to 0 \text{ and } \lim_{T\to\infty} \mu(T)S(T)\to 0.$$

Now equations (5.5) and (5.7) yield the Ramsey savings rule. μ in equation (5.6) is negative, being the difference between the price of land in forestry and in agriculture. This can be seen more clearly if we integrate equation (5.8) to get $\mu(t)$, the util-valued wedge in land prices. From equation (5.8) we have:

$$-\mu(t) = \int_t^\infty e^{\rho(v-t)}\lambda(V)\phi(S(v))Q_L(v)\mathrm{d}v - \int_t^\infty e^{-\rho(v-t)}\lambda(v)\phi_S(S(v))Q(v)\mathrm{d}v.$$

The first term on the right-hand side of this equation is the util-valued present value of a stream of marginal products of land in agriculture. We label this price $\lambda(t)P_L(t)$. This is the price (capital value) of a unit of land in agriculture, valued in utils. The last term on the right-hand side is the present value of a stream of marginal products of biodiversity (the marginal value of an extra unit of land in forest). We label this price $\lambda(t)P_s(t)$. Using this in equation (5.6) yields:

$$P_L(t) - P_S(t) = g_R(t)$$

or the wedge in land prices is the marginal cost of transforming a unit of cost of land in forest to a unit in agriculture. The subtle point is that $P_S(t)$ will be set at zero in a free market situation because the value of land in forest from a free market standpoint is simply as a reservoir for agricultural land. Only in unusual circumstances will land in forest be valued as well for its capacity to sustain biodiversity. This is the *indirect benefit* or *value* of a unit of land in forest. The direct benefit here is simply to be available to be transformed into agricultural land. $P_S(t)$ is then *Pigovian price* or tax on forested land which must be paid once-over before a unit can be cleared for agricultural use. This makes the cost of a unit of agricultural land to a producer, $P_S(t) + g_R(t)$ and not simply $g_R(t)$. A Pigovian tax internalizes the unpriced marginal value of forest in increasing biodiversity, $S(t)$. The rental payment per period that society owes to forest owners in preserving biodiversity is $\phi_S(S(t))Q(t)$. Discounted rentals constitute the price of a unit of land in forest. Pearce (1992) suggested instituting an annual payment of about

$(US) 3.5 billion to owners of forested land in the Amazon in order to 'compensate' them for providing indirect services (non-marketed services) such as providing a place for biodiversity to be sustained. He was valuing a range of indirect services, not just the sustaining of biodiversity.

If we substitute from equations (5.4) and (5.6) into (5.3), we get the dollar value net national product, NNP(t), for this economy:

$$\frac{H}{U_C} = \frac{U(C)}{U_C} + \dot{K} + Rg_R. \tag{5.9}$$

Equation (5.9) indicates that the change in the value of land, g_R, and the amount of land deforested, $R(t)$, show up as an upward capital valuation in NNP. Positive values of g_R and R imply that land in agriculture is more valuable than land in forest. That biodiversity is valuable shows up in $P_S(t)$ being positive. Given Pigovian tax P_S, the loss of biodiversity will be slowed because deforestation is more expensive than it would be with a zero price for the indirect benefits from forests, namely preserving biodiversity. But P_S appears in the NNP function as $(P_L - P_S)R_t$ (equal to $g_R R_t$) and in a sense is rendered immaterial for our purposes because it enters as part of a price change and not as a price level. We are accustomed to valuing declining natural stocks in NNP as a price level (marginal value) multiplied by a change in stock (as suggested by Hartwick (1990)). Here, there is indeed a change in stock, namely R hectares being taken out of forest and into agriculture, but the valuation is in terms of a price change or a marginal cost of changing land use. Thus Pigovian taxes show up in NNP only indirectly and we see a case where capital gains should be included in NNP. (Eisner (1988) advocated introducing land 'revaluation' in NNP(t). Note here that the capital gain is only on land whose use is changed, not on all land.)

Model II: Pigovian tax on land being deforested

Now we consider a discrete once-over decline in the efficiency parameter ϕ as biodiversity declines. The date of the decline is a random variable, depending on the size of the loss in biodiversity. $F(L(t))$ is the probability of the decline occurring beyond time t and $f(L(t))(R(t)$ is the probability of the decline occurring at t. In this situation, the expected present value of welfare is (see Appendix 1 of this chapter for details):

$$\Omega(t) = \int_0^\infty e^{-\rho t} U(C)F(L(t))\mathrm{d}t + \int_0^\infty e^{-\rho t} \rho W[K(t), S(t)]f(L(t))R(t)\mathrm{d}t,$$

where:

$$W(K(t), S(t)) = \int_{t}^{\infty} e^{-\rho(v-t)} U(C^*(v)) dv,$$

C^* indicating the optimal value of $C(t)$, given initial stocks $K(t)$ and $S(t)$. Thus, after ϕ declines discretely to $\underline{\phi}$, there is no further consequence of the presumably continued loss in biodiversity. Before the decline, ϕ is at $\bar{\phi}(>\underline{\phi})$ and the date of the decline is a random variable increasing with t (and the loss in biodiversity). The equations of motion up to t are $\dot{K} = \bar{\phi}Q(K, N, L) - g(R) - C$ and $\dot{S} = -R$. Beyond t, we have $\dot{K} = \underline{\phi}Q(K, N, L) - g(R) - C$ and $\dot{S} = -R$. (But, beyond t, the outcome of our economy growing is captured in $W(K(t), S(t))$.) The current-value Hamiltonian for this problem is:

$$H = U(C)F(L) + \rho W(K, S)f(L)R + \lambda(t)[\bar{\phi}Q(K, N, L) - g(R) - C] - \mu(t)R.$$
$$(5.10)$$

Necessary conditions for a maximum are:

$$\frac{\partial H}{\partial C} = 0 \rightarrow U_C F = \lambda, \qquad (5.11)$$

and

$$\frac{\dot{\lambda}}{\lambda} = \frac{\dot{\overline{U_C F}}}{U_C F}, \qquad (5.12)$$

$$\frac{\partial H}{\partial R} = 0 \rightarrow \rho Wf - \lambda g_R = \mu, \qquad (5.13)$$

$$-\frac{\partial H}{\partial K} = \dot{\lambda} - \rho\lambda \rightarrow \frac{\dot{\lambda}}{\lambda} = \rho - \bar{\phi}F_K - fR\rho W_K, \qquad (5.14)$$

$$-\frac{\partial H}{\partial S} = \dot{\mu} - \mu\rho \rightarrow \frac{\dot{\mu}}{\mu} = \rho + \frac{U(C)F_L}{\mu} - \frac{fR\rho W_S}{\mu} + \frac{\rho WRf_L}{\mu} + \frac{\lambda\bar{\phi}Q_L}{\mu} \quad (5.15)$$

The central result in equations (5.11) to (5.15) is (5.13). ρWf is the Pigovian tax on land currently deforested. The private marginal benefit (capital gain) of transferring a hectare from forest to agriculture is g_R. The social marginal benefit is $g_R - \rho Wf/\lambda$, rather less because deforestation diminishes biodiversity and in so doing increases the probability of ecological collapse in the near term. The social marginal benefit reflects the marginal value of biodiversity in forestalling ecological collapse. Expression $-\mu/\lambda$ is the marginal social payoff from moving a hectare from forestry to agriculture. If this declines to zero, then $R(t)$ declines to zero in the positive land transfer tax regime. This is a boundary solution in our model.[3] Clearly $R(t) = 0$ is a species of sustainable development.

Observe then that $fR\rho W_K$ is the change in tax revenue as K changes. The tax rate $f\rho W$ varies over time. $fR\rho W_S$ is the change in tax revenue as S changes. W_S will be negative as long as $R(t) > 0$. This reflects the fact that an extra hectare in forestry is worth less than an extra hectare in agriculture as long as deforestation is continuing.

The combination of equations (5.12) and (5.14) becomes the Ramsey optimal savings rule for this model. Equation (5.13), its derivative with respect to time, and equation (5.15) yield the 'Hotelling rule' for this model. Of considerable interest is the expression for NNP(t) in this model. If we substitute for λ and μ from equations (5.11) and (5.13) into (5.10), we obtain the util-valued NNP(t). That is:

$$H = F(L)\left\{ U(C) + U_C \dot{K} + \left[U_C g_R R - \frac{\rho W U_C fR}{U_C F} \right] + \frac{\rho W fR}{F} \right\}. \quad (5.16)$$

The term in square brackets is the social capital gain (in utils) on land, R, transferred from forest to agriculture. $\rho W fR/U_C F$ is the total Pigovian tax on the land – this tax internalizes the external effect of deforestation on the probability of ecological collapse. The expression in curly brackets is net marginal social payoff in utils. The expression:

$$U(C) + U_C \dot{K} + \left[U_C g_R - \frac{\rho W f U_C}{U_C F} \right] R$$

is a standard NNP(t) measure in utils at date t. The term on the end in equation (5.16) has hazard rate, fR/F, multiplied by capital value W. This last term is a risk adjustment. Full NNP(t) at t comprises a scarcity value NNP plus a risk premium reflecting the possibility of collapse at t. The hazard rate is the conditional probability of collapse at t, given no collapse up to t. $F(L)$ in front of the braces is the probability of the state at t being 'no collapse yet'.

The striking result in equation (5.16) is the risk premium being equal to the current Pigovian taxes paid essentially to reduce the risk of collapse at

---

[3] A formal treatment of the model evolving to a boundary solution and beyond is an open research question.

any date, marginally.[4] The effect of the Pigovian taxes is in fact to delay marginally the expected date of collapse.[5] The marginal (and total) cost of 'reducing' the risk of collapse at t equals the marginal (and total) benefit of 'reducing' risk. The linearity of marginal cost and marginal benefit in terms of R makes total benefits and costs equal. The benefits of 'reducing' the risk of collapse at t are fully capitalized in the costs of 'reducing' risk.

The dollar valued NNP(t) is:

$$\frac{H}{U_C} = F(L)\left\{ \frac{U(C)}{U_C} + \dot{K} + \left[g_R - \frac{\rho Wf}{U_C F} \right] R + \frac{\rho WfR}{U_C F} \right\}. \qquad (5.17)$$

It is appropriate to refer to H/U_C as *ex ante* dollar-valued NNP(t) at t and the expression in braces as *ex post*, since the term in curly brackets is the NNP at t realized if collapse has not occurred up to date t. $F(L)$ is the probability of no collapse by t. Note $F(L)$ will be approximately unity early in the program and will decline slowly toward zero as time passes. Thus the gap between *ex ante* and *ex post* NNP will increase up to the date of collapse. After collapse, there is no further risk of collapse and *ex ante* and *ex post* values will be the same.

Weitzman (1976, p. 161) distinguished between anticipated technical changes of a 'sporadic or stochastic' nature that would be ultimately capitalized in increments to the value of capital stocks, including stocks of knowledge and unanticipated technical change. 'The law of large numbers' was appealed to with regard to anticipated technical change. There is seemingly implicit either a replacement of random variables with their corresponding expected values or valuation in *ex ante* terms. Unanticipated shocks were viewed as a once-over jump in current NNP(t). These examples of anticipated structural breaks suggest that a separate analysis is needed for cases of anticipated shocks when the law of large numbers does not apply. The NNP(t) value here has a distinct risk term and a wedge between *ex ante* and *ex post* NNP(t). Weitzman's analysis does not cover this case.

Concluding remarks

We have seen two sorts of Pigovian taxes required to internalize the indirect cost of a diminution in biodiversity. In the case of no structural break, the

---

[4] A similar cancellation appeared in an analysis of durable exhaustible resources and NNP by Hartwick (1991a). (See also Hung, 1993.) Another such cancellation occurs in the analysis of deforestation and NNP by Hartwick (1992). In Appendix 2, this same cancellation can be observed in Heal's model of carbon emissions and hydrocarbon mining.

[5] Paying to alter the expected date of an event is focused on by Hartwick (1991a), a discussion of rewards in research and development races.

'tax' turned out to be a charge on each unit of land deforested. But this charge did not enter into NNP directly. We seemingly have an unadjusted NNP(t) 'formula' with indirect costs being internalized. The 'tax' enters into the formula for capital gains on land deforested but not into the level of capital gains. In our second model the gradual loss in biodiversity caused the probability of ecological collapse to gradually increase. This increase in probability was internalized by a Pigovian tax on land currently deforested. At each date the current risk of collapse showed up as a risk factor term in a risk-adjusted NNP 'formula'.

Appendix 1: The objective function when collapse is uncertain

We equate our index of biodiversity, $S(t)$, with land, $S(t)$, remaining in forest. Then land deforested is $S_0 - S(t) (= L(t))$. We define the probability that collapse occurs between t_1 and t_2 as:

$$\text{Prob}\,(T \in (t_1, t_2)) = \int_{L^{t_1}}^{L_{t_2}} f(L_t)\mathrm{d}t.$$

Also the probability of collapse at T or beyond T is:

$$F(L_t) = \int_T^\infty f(L_t)R_t\mathrm{d}t,$$

where $R_t = \dot{L}_t$. The maximand (an expected value) is:

$$E\left\{\int_0^T U(C_t)e^{-\rho t}\mathrm{d}t + W(K_T, S_T)e^{-\rho T}\right\},$$

$$= \int_{L_0}^{S_0} f(L)\left\{\int_0^T U(C_t)e^{-\rho t}\mathrm{d}t + W(K_T, S_T)e^{-\rho T}\right\}\mathrm{d}L,$$

$$= \int_0^\infty f(L)R\left\{\int_0^T U(C_t)e^{-\rho t}\mathrm{d}t + W(K_T, S_T)e^{-\rho T}\right\}\mathrm{d}T,$$

$$= \int_0^\infty U(C_t)F(L_t)e^{-\rho t}\mathrm{d}t + \int_0^\infty \rho W(K_t, S_t)f(L_t)R_t e^{-\rho t}\mathrm{d}t.$$

Appendix 2: Carbon taxes and risk-adjusted NNP

Heal's (1984) model of once-over uncertain collapse of an economy because of a build-up of CO_2 contains implicitly both a Pigovian carbon tax, a risk-adjusted $NNP(t)$, and the exact capitalization of current risk in current carbon taxes. Carbon is released as hydrocarbons are mined. $R(t)$ tons are mined at date t leaving $S(t)$ tons remaining. Cumulative extraction $Z(t) = S_0 - S(t)$. The probability of collapse ($\bar{\phi}$ jumping down to ϕ) in interval (t_1, t_2) is $\int_{z(t_1)}^{z(t_2)} f(Z) dZ$. Thus, the probability of collapse at date t rises with cumulative extraction $Z(t)$. $R(t)$, capital $K(t)$, and labor N (constant) are used to produce output $Q(t)$ in $Q(K(t), N, R(t), \bar{\phi})$. The equations of motion are:

$$\dot{K}(t) = Q(K(t), N, R(t), \bar{\phi}) - C, \tag{5.18}$$

$$\dot{S}(t) = -R(t). \tag{5.19}$$

We have initial stocks K_0 and S_0. After the low level state ϕ is realized at date T, and $K(T)$ and $S(T)$ remain, realized welfare beyond T is $W(K(T), S(T)) = \int_T^\infty e^{-\rho t} U(C^*(t)) dt$. Expected discounted $U(C(t))$ is maximized where the random variable is the date of collapse t. The current-value Hamiltonian for this problem is:

$$H(t) = U(C)F(Z(t)) + \rho W(K(t), S(t)) f(Z(t)) R(t)$$
$$+ \lambda(t)[Q(K(t), N, R(t), \bar{\phi}) - C(t)] - \mu(t) R(t), \tag{5.20}$$

where $F(Z(t))$ is the probability of collapse beyond date t, and $f(Z(t)) R(t)$ is the probability of collapse at t. The conditional probability of collapse at t, given no collapse up to t, is $f(Z(t)) R(t)/F(Z(t))$, i.e. the hazard rate at t. $\lambda(t)$ is the util-valued price (capital value) of a unit of $K(t)$ and $\mu(t)$ is the util-valued price of a unit of $S(t)$. Since $R(t)$ extracted is immediately consumed, $\mu(t)$ appears in equation (5.23) as a flow price rather than a capital value. Necessary conditions for an optimum are:

$$\frac{\partial H}{\partial C} = 0 \rightarrow FU_C = \lambda, \tag{5.21}$$

and:

$$\frac{\dot{\lambda}}{\lambda} = \frac{\dot{U}_C}{U_C}, \tag{5.22}$$

$$\frac{\partial H}{\partial R} = 0 \rightarrow \rho Wf + \lambda Q_R = \mu(t), \tag{5.23}$$

$$-\frac{\partial H}{\partial K}=\lambda-\rho\lambda \;\rightarrow\; -\rho W_K fR-\lambda Q_K=\lambda-\rho\lambda, \qquad (5.24)$$

$$-\frac{\partial H}{\partial S}=\mu-\rho\mu \;\rightarrow\; U(C)F_Z+\rho WRf_Z-\rho fRW_S=\mu-\rho\mu. \qquad (5.25)$$

Equations (5.18) to (5.25) may be found in the article by Heal (1984). The principal economic result is that $\rho Wf/\lambda$ is a Pigovian (carbon) tax in equation (5.23) on mining of $R(t)$ of hydrocarbons. This tax internalizes the spillover of increased risk of collapse at t from marginally more mining. Mining $R(t)$ has a productivity effect Q_R and a side effect Wf/λ; the dollar social price of a unit of $R(t)$ is $\mu(t)/\lambda(t)=Q_R+\rho Wf/\lambda$. The private price Q_R fails to ration extraction properly because it neglects the external effect of cumulative extraction contributing to increased probability of $\bar{\phi}$ shifting down to ϕ.

Equations (5.21), (5.22) and (5.24) yield the risk-adjusted Ramsey savings rule for this economy. Equations (5.21), (5.23) and (5.25) yield the risk-adjusted Hotelling rule for extracting $R(t)$ at each date before collapse. If we substitute for $\lambda(t)$ from equation (5.21) and $\mu(t)$ from equation (5.23) into (5.20), we get the util-valued NNP(t) function:

$$H(t)=F(Z(t))\left[U(C)+\dot{K}U_C-\left\{U_C Q_R+\frac{\rho Wf}{F}\right\}R+\frac{\rho WfR}{F}\right]. \qquad (5.26)$$

The last term in equation (5.26), $\rho WfR/F$ is a risk factor, the conditional probability of collapse at t, multiplied by the capital value of the program after collapse. The first term $F(Z(t))$ is roughly speaking an index of the state of the system. (Formally $F(Z(t))$ is the probability of collapse beyond t.) $(\rho Wf/F)R$ in $\{...\}R$ in equation (5.26) is the Pigovian tax internalizing the risk-of-collapse spillover. The expression $U(C)+\dot{K}U_C-\{...\}R$ is a standard NNP expression for an economy without risk of collapse. This implies that $[...]$ in equation (5.26) is a risk-adjusted NNP expression. True *ex post* NNP before collapse, namely $[...]$ in equation (5.26), contains a risk term. *Ex ante* NNP before collapse is $F(Z(t))[...]$. $F(Z(t))$ is approximately unity early in the program for $Z(t)$ small and tends to zero as $Z(t)$ tends to S_0. Thus the gap between *ex ante* NNP and *ex post* NNP (risk adjusted *ex post* NNP) increases as time passes.

Risk $\rho WfR/F$ in equation (5.26) is fully capitalized in Pigovian taxes. We observed this result above in the text in our loss-of-biodiversity model. It seems to be an artifact of the linearity of the expectation operation.

References

Cropper, M. L. (1976). Regulating activities with catastrophic environmental effects. *Journal of Environmental Economics and Management*, **3**, 1–15.

Eisner, R. (1988). Extended accounts for national income and product. *Journal of Economic Literature*, **26**, 1611–84.

Hartwick, J. M. (1990). Natural resources, national accounting and economic depreciation. *Journal of Public Economics* **43**, 291–304.

Hartwick, J. M. (1991a). Notes on economic depreciation of natural resource stocks and national accounting. In A. Franz, C. Stahmer, and S. Fickl, eds. *Approaches to Environmental Accounting*, pp. 167–219. Physica Verlag, Heidelberg.

Hartwick, J. M. (1991b). Patent races optimal with respect to entry. *International Journal of Industrial Organization*, **9**, 197–207.

Hartwick, J. M. (1992). Deforestation and national accounting. *Environmental and Resource Economics*, **2**, 513–21.

Heal, G. (1984). Interactions between economy and climate: a framework for policy design under uncertainty. In V. Kerry Smith and A. D. Witte, eds. *Advances in Applied Micro-Economics*, vol. 3, pp. 151–68. JAI Press, New York.

Hung, N. H. (1993). Natural resources, national accounting, and economic depreciation: stock effects. *Journal of Public Economics*, **51**, 379–90.

Pearce, D. (1992). Deforesting the Amazon: toward an economic solution. *Ecodecision*, **1**, 40–50.

Weitzman, M. L. (1976). On the welfare significance of national product in a dynamic economy. *Quarterly Journal of Economics*, **90**, 156–62.

6

Biodiversity conservation as insurance

CHARLES PERRINGS

Introduction

While most of the debate about the significance of biodiversity loss has centred on genetic information lost through species extinction, recent research has switched the focus from the characteristics of particular organisms to the functionality of the mix of organisms in ecosystems (Holling *et al.*, 1994). The 'cure for cancer' has been displaced by the role played by the mix of species and communities in maintaining the resilience of ecosystems. This has important implications both for the way we think about the social costs of biodiversity loss and for the efficacy of policy instruments designed to deal with the problem. By changing our perception of the nature of the social costs of biodiversity depletion, the link that is now being emphasised between functional diversity and ecological resilience also changes our perception of the effectiveness with which the problem may be addressed at the global level. This is because it changes both the time path and the geographical distribution of the benefits of biodiversity conservation. The effectiveness of biodiversity conservation policy at the global level is a function of the geographical distribution of benefits.

More particularly, if the main short- and medium-term implications of biodiversity depletion lie in the genetic information losses associated with species or population deletion, then it follows that the main short- and medium-term social costs of biodiversity loss are global, and the probability of developing an effective biodiversity strategy are reduced. But if the main short- and medium-term implications of biodiversity depletion lie in the loss of ecosystem resilience, then the main social costs of biodiversity loss will be ecosystem specific, and the probability of developing an effective biodiversity strategy are increased. Put another way, if a large part of the social costs of biodiversity are local, not only is the problem easier to address within the nation state, but the relative value of the transfers needed to induce international cooperation is reduced. Species deletion at private hands may still have implications for the global community, but, where the main external costs of private actions leading to loss of functional diversity accrues to other users of the same ecosystem, there is a strong incentive for local governments to address the problem directly.

This chapter considers the social benefits of biodiversity conservation from this perspective. More particularly, it considers the insurance value of conservation at the national and international levels. The chapter is in five sections. The second section discusses the ecological arguments for the insurance value of biodiversity conservation in the context of the recent focus on the role of biodiversity in maintaining ecosystem resilience. The third section then discusses the insurance value of biodiversity conservation. The fourth section discusses the policy implications of the approach, and a final section offers some concluding remarks.

Biodiversity as 'insurance' in ecological systems

Insurance is conventionally conceptualised as a means of pooling actuarial 'risk'. While the 'risks' associated with changes in the organisation of jointly determined ecological–economic systems may be computable in an actuarial sense in some cases, in many more cases they will not. In many cases neither the set of outcomes of a change in the organisation of such systems nor the probability of occurrence of each outcome will be known. In these circumstances, the problem is not one of risk, but of fundamental uncertainty. The insurance value of biodiversity conservation concerns its role in the amelioration of fundamental uncertainty, rather than risk.

To begin with, it is important to make one general observation about the dynamics of ecological systems. In all non-linear ecological systems the system dynamics are characterised by the existence of multiple locally stable equilibria (or basins of attraction), separated by unstable equilibria (or unstable manifolds) that are defined in terms of the level or density of the state variables or components of the system. A change in the organisation of the system implies a change in the system equilibrium or attractor. This implies that the system dynamics may be neither continuous nor gradual, and that changes inducing a switch from one basin of attraction to another may be irreversible. The system's response to perturbation depends both on where it is relative to the system equilibria and on the characteristics of those equilibria. So, if a system is in the neighbourhood of a particular unstable equilibrium, or threshold, minor perturbation of its state variables may have 'catastrophic' consequences for its structure and organisation. Conversely, if a system is at or close to a locally stable equilibrium, major perturbation of the same variables may have very little effect on its structure or organisation.

In terrestrial ecology, this has stimulated an approach to the analysis of system dynamics that concentrates on where a far-from-equilibrium ecosystem is relative to the unstable manifolds or thresholds of the general system. The approach requires identification not of the existence and stability of equilibria, but of resilience: the capacity of a system to absorb shocks without losing stability (Walker and Noy-Meir, 1982; Holling, 1973; Schindler, 1990). In terrestrial systems threshold values exist for a wide range of ecological functions. If these values are exceeded, the system loses resilience, or the ability to maintain its self-organisation without undergoing the 'catastrophic' and irreversible change involved in crossing such thresholds. Resilience is therefore a measure of the magnitude of disturbance that can be absorbed before the system flips from one basin of attraction to another. For the purposes of this chapter, it is also a measure of the magnitude of disturbance that can be sustained before the system loses predictability.

This follows from the fact that the level of risk and fundamental uncertainty associated with the future dynamics of the system is a function of where the system is relative to the thresholds of instability (unstable manifolds). A system close to the unstable region that marks the basin boundary (meaning that its resilience with respect to perturbation in the direction of the basin boundary is low) may be dislodged from the basin by a minor perturbation. If it flips from a familiar to an unfamiliar basin, its dynamics will in general be fundamentally uncertain. That is to say, the future states of nature associated with that perturbation and the probability of occurrence of those states will be unknown.

The role of biodiversity in this lies in its link with system resilience. Systems ecologists now take the view that the dynamics of most terrestrial ecosystems are dominated by a small set of processes (Holling, 1992), and that the dynamics of species may be more sensitive to ecosystem stress than the dynamics of processes (Schindler, 1990; Vitousek, 1990). In other words, stressed ecosystems may maintain many of their functions even though the composition of the species comprising those ecosystems changes. However, the ability of the key structuring processes of a system to operate under a range of conditions depends on the number of alternative species that can take over functions when perturbation of an ecosystem causes the disappearance of the species currently supporting those functions (Schindler, 1988). In short, it is the functional diversity of ecosystems that determines their resilience. Indeed, there is growing evidence that the least resilient components of food webs, energy flows and biogeochemical cycles are those in which the number of species carrying out important functions is very small (Schindler, 1990).

The link between biodiversity, ecosystem resilience and insurance should now be transparent. Other things being equal, the greater the mix of species in terrestrial systems, the greater the resilience of those systems implying the greater the perturbation they can withstand without losing their self-organisation. Biodiversity underpins the ability of far-from-equilibrium ecological systems to function under stress, and in so doing it underpins the predictability of those systems. Greater levels of biodiversity protect the system from the frequently unpredictable and irreversible effects of the change in self-organisation associated with change in attractor or equilibrium state. It follows that the value of biodiversity conservation lies in the value of that protection: the insurance it offers against catastrophic change.

The value of insurance in ecological–economic systems

The context in which biodiversity conservation strategies are currently being developed is not, of course, the time behaviour of untouched natural systems, it is the time behaviour of jointly determined ecological–economic systems. In all such systems, organisational change generates two inter-linked sets of 'general equilibrium' effects: a set of ecological effects that work themselves out in the evolution of the ecological systems concerned, and a set of economic effects that work themselves out in the evolution of the economic system. The cross-effects depend on the 'connectedness' of the two systems. The more highly connected are the ecological and economic systems, the more change in one implies change in the other. But there is both a spatial and a temporal structure to the connections between the economy and its environment. Hence components of the joint system may be entirely unconnected from the perspective of one time horizon or one geographical region, but may be highly connected from the perspective of some other time horizon or geographical region.

The problem for policy makers is that the market prices that are the principle observers of the joint system are, in these circumstances, very poor indicators of the opportunity cost of committing particular classes of resource to some economic use. Many of the reasons why prices are inadequate observers of ecological resources are well understood. They include a lack of understanding of the system dynamics, the structure of property rights, the effects of government policy, the public good nature of some ecological resources and so on. These are discussed elsewhere in this volume and need not detain us here. The net effect is that the market prices of ecological resources cannot register the true costs of their allocation in the neighbourhood of system thresholds. If there exist multiple equilibria, most of which have not been observed, there will be a very large measure of fundamental uncertainty about the future effects of current actions wherever these threaten to cross critical ecological thresholds. It will not be possible to calculate the actuarial value of the environmental 'risks' involved wherever these 'risks' include a loss of system resilience.

Allocation of ecological resources on the basis of market prices will be less efficient the closer the joint system is to critical thresholds. At the same time, estimation of the social opportunity cost of the allocation of ecological resources will be more difficult. In other words, the closer the joint system is to thresholds, the more will the private cost of ecological resources allocated to some use tend to understate the cost to society, but also the more difficult will it be to estimate that cost.

It is possible to approximate the 'premium' paid for insurance against the fundamentally uncertain implications of loss of resilience. Consider two cases. The first is the case of intensive agriculture. The dominant characteristic of intensive agriculture is a reduction of biodiversity in order to focus on particular species with properties that include, *inter alia*, high rates of growth, i.e. the crops selected in intensive agriculture are generally r-strategists with high rates of net primary productivity. This typically leads to an overall reduction in the resilience of the ecological systems on which intensive agriculture is based, and to the growing dependence of agriculturists on intensive managment regimes in which output is frequently maintained only by ever-increasing use of fertilisers, pesticides and irrigation. Agricultural systems have to be ever more tightly protected against perturbation (disease, drought and so on) precisely because their sensitivity to perturbation has increased with specialisation.

Without external access to substitute plants and genetic material for engineering disease and pest resistance, and without the input of biocides the approach is extraordinarily susceptible to perturbation. The cost of the management regime needed to maintain a system of very low natural resilience provides a measure of the benefits of ecosystem resilience. Where intensive management has substituted for diversity of species as the primary insurance against collapse of the agricultural system, then the cost of the management regime may be thought of as the insurance 'premium' on system resilience.

The second case is that of regulated-access fisheries. Individual fishermen are enjoined to avoid driving particular species to extinction, on pain of some penalty no less than the maximum private benefit to be had from so doing. In terms of a simple population indicator, for example, if the growth function for some population is critically depensatory, assurance against extinction of that population may be provided through the penalties imposed on users depleting the population past the critically depensatory point. In practical terms this might translate as a combination of catch quota and gear restrictions. The implicit social insurance premium would then be the sum of the benefits forgone by limiting the size of the harvest. Of course, if quotas are set at levels that do not protect the populations being exploited, the implicit social insurance premium is zero. That is to say, no social insurance has been taken out against the 'risk' of population collapse.

Neither case provides a measure of the premium on biodiversity per se, but both offer an indication of the premium on maintaining resilience under a given allocation of resources. In principle, however, wherever the conservation of biodiversity does have an opportunity cost in terms of output, the aggregate value of output forgone provides a first approximation of the value to the conserving authority of the maintenance of ecosystem resilience through conservation. The problem for international policy posed by the loss of biodiversity is that there may be no reason to believe that this first approximation is a measure of the social opportunity cost of loss of system resilience.

The distribution of the benefits of biodiversity conservation

A generally recognised difficulty in devising policies to protect the resilience of systems where this is sensitive to the mix of species is due to the fact that the ecological services of biodiversity are in the nature of a 'layered' public good. Biodiversity conservation yields benefits at several different levels. The benefits of biodiversity conservation in terms of the genetic library are non-exclusive at the global level. Similarly, the benefits of biodiversity in terms of local ecosystem resilience are non-exclusive at the local or ecosystem level. But individual populations are both rival and exclusive in consumption, and occur in multiple jurisdictions. In other words, the benefits of biodiversity conservation are diffused across international boundaries, whereas the benefits of hunting or harvesting are captured by the individual resource user.

It follows that the insurance taken out by the user will generally be less than the globally optimal level of insurance, implying that the 'functional redundancy' in the ecological system will be less than the social optimum.

Put another way, there will be insufficient resources committed to what may be termed 'ecological stabilisation': the maintenance of the productive potential of ecosystems supplying essential ecological services either by the ontainment of stress levels or by the promotion of ecosystem resilience through biodiversity conservation. This said, if the ecologists are correct that the major part of the benefits of biodiversity conservation lie in the maintenance of the resilience of locally exploited ecosystems, at least in the short and medium term, then the gap between the national and global benefits of conservation may not be as large as has been thought. The distribution of the global benefits of biodiversity conservation matters.

The aim of an ecological stabilisation strategy is either to restrict the level of stress on ecosystems to levels at which they are resilient, or to increase the resilience of ecosystems by appropriate intervention, or both. If the level of stress on any system is such that the system is not resilient, then the imposition of quotas on one or more commercial species will merely ensure that it is the collapse of some other species that causes the system to lose self-organisation. It turns out, however, that regulating the general level of pressure, rather than pressure on particular points of a system, is not as difficult as it seems. If the resource ultimately being exploited is not some population, but the system that supports that population, then it is reasonable for society to restrict access to the system, and to charge a royalty for access. This has the effect of converting a public good into a 'club good', whilst simultaneously establishing a premium for insuring against the fundamental uncertainty associated with the collapse of the system. Indeed, regulation of the general level of stress on the system is a key feature of the management of uncertainty. Restricting access to the system protects its resilience, which in turn protects against the unanticipatable outcomes of collapse of the self-organisation of the system.

The global interest in the effectiveness of local measures for the management of fundamental uncertainty requires 'side payments' to adjust the set of payoffs associated with strategies that do and do not protect system resilience. The appropriate strategy is one that uses transfers (to finance the conservation of biodiversity) in a way that ensures that the payoff to conservation dominates the payoff to non-conservation. The transfers in the fisheries case just discussed, for example, might be used to 'buy out' those with historically free rights of access to stressed systems, but the precise form that they take might be expected to differ from system to system. In general, however, the smaller the difference between the cooperative and non-cooperative outcomes, the greater the probability that there will exist a set of transfers that supports the cooperative outcome.

Barrett (1994) considered the conditions in which the transfers intended to meet 'incremental costs' (the difference in payoffs between cooperative and non-cooperative strategies) under the biodiversity convention might be self-enforcing. He argued that agreement can sustain the full cooperative outcome only where the global net benefits of cooperative behaviour are 'slightly larger' than the global net benefits of non-cooperative behaviour. It follows that if the global externalities of biodiversity loss are very much larger than the local externalities, then biological diversity convention cannot offer a solution to the problem. However, if the difference is not large – as the work on the role of biodiversity in ecosystem resilience implies – the prospects for the biodiversity convention are much brighter.

Concluding remarks

To return to the problem of the insurance value of biodiversity, it is axiomatic that environmental effects that are in the nature of actuarial risks may be insured against on a commercial basis. Indeed, for all such risks, the appropriate strategy is for the environmental authority to require private users of environmental resources to take out cover against potential losses associated with their actions. A difficulty arises only where the economic use of ecological resources involves fundamental uncertainty: i.e. where it involves 'risks' that are not commercially insurable. In such cases, the appropriate strategy is one of ecological stabilisation, requiring restriction of the level of stress on the system concerned or an increase in the capacity of that system to respond creatively to both stress and shock.

Biodiversity conservation is, from this perspective, a means of managing the uncertainty that lies beyond the basin boundaries of the joint ecological–economic system. It also protects the set of choices bequeathed to future generations. If an agent has the power to choose from some opportunity set, then he or she cannot lose by adding an alternative to that opportunity set, even if that alternative is strictly worse than existing alternatives (Perrings, 1994). This means that neither the present nor future generations can be made worse off if the present generation retains an alternative (some species of plant or animal) even if it is strictly worse than all other alternatives under the preference ordering of the present generation. This may be regarded as an incidental benefit of the management of uncertainty through biodiversity conservation, but it is not a trivial one.

References

Barrett, S. (1994). The biodiversity supergame. *Environmental and Resource Economics*, **4**, 111–22.

Holling, C. S. (1973). Resilience and stability of ecological systems. *Annual Review of Ecology and Systematics*, **4**, 1–23.

Holling, C. S. (1992). Cross-scale morphology geometry and dynamics of ecosystems. *Ecological Monographs*, **62**, 447–502.

Holling, C. S., Schindler, D. W., Walker, B. W. and Roughgarden, J. (1994). Biodiversity in the functioning of ecosystems. In C. Perrings, C. Folke, C. S. Holling, B. O. Jansson and K. G. Mäler, eds. *Biological Diversity: Economic and Ecological Issues*, pp. 44–83. Cambridge University Press, New York.

Perrings, C. (1994) Biotic diversity, sustainable development and natural capital. In A. M. Jansson, C. Folke, R. Costanza and M. Hammer, eds., *Investing in Natural Capital*, Island Press, Covelo (in press).

Schindler, D. W. (1988). Experimental studies of chemical stressors on whole lake ecosystems. Baldi Lecture. *Verhandlungen Internationale der Verein bei Limnologen*, **23**, 11–41.

Schindler, D. W. (1990). Experimental perturbations of whole lakes as tests of hypotheses concerning ecosystem structure and function. Proceedings of 1987 Craoford Symposium. *Oikos*, **57**, 25–41.

Vitousek, P. M. (1990). Biological invasions and ecosystem processes: towards an integration of population biology and ecosystem studies. *Oikos*, **57**, 7–13.

Walker, B. H. and Noy-Meir, I. (1982). Aspects of the stability and resilience of savanna ecosystems. In B. J. Huntley and B. H. Walker, eds. *Ecology of Tropical Savannas*, pp. 577–90. Springer-Verlag, Berlin.

7

Property rights, externalities and biodiversity

ROGER A. SEDJO AND R. DAVID SIMPSON

It is well recognized among economists that property rights play a critical role in the efficient utilization of natural resources. The absence of property rights to resources results in systematic overexploitation of the resource, sometimes referred to as the 'tragedy of the commons'. In contemporaneous society the broad social concern over the decline in biological diversity can be viewed by the economist as a concern over social values lost, i.e. negative externalities, that result from the biodiversity decline. Since the resource is owned by no one, the financial incentives are for short-term exploitation, because there is no certainty of capturing long-term returns. Thus, there are no financial incentives for long-term protection and development of the resource.

An institutional approach to reducing externalities associated with biodiversity can be found in the creation of 'rights' to genetic resources whereby an owner-agent can capture the benefits over the long-term and therefore has an incentive to provide resource protection and development. To fail to do so results in financial loss through the deterioration or destruction of resource values. Thus, an incentive for protection of the resource as a financial asset is created. This approach is discussed here.

The question arises as to what is the nature of the values of wild genetic resources and what are the social losses associated with the decline of biodiversity. A ready response is that these resources have direct value in use as in natural products or as inputs into the process of creating new drugs, medicines and pharmaceutical products. The value to society of wild genetic resources is often discussed in terms of these resources potentially holding natural compounds that offer, say, a cure for cancer or AIDS.

In addition to use values, there appears to be a spectrum of other values attributed to biodiversity. These run from values that clearly are heartfelt to some, and often difficult to articulate or describe with any great precision, to values believed to be associated with system stability, but lacking a compelling scientific basis. These non-use values cover the range from spiritual-like beliefs that natural biological systems ought not be substantially disturbed, to a belief that the biological system is potentially highly unstable and disturbances could forever alter it in some unspecified manner that would have substantial negative consequences for humans. Economists have named some of these values: existence value or non-use values; option values, associated with uncertainty regarding future use; and quasi-option values, reflecting intertemporal aspects of irreversible decisions.

The focus of this chapter is on the use value of genetic resources, specifically the value of wild genetic resources as inputs into pharmaceuticals and natural products. First, we present an overview that defines and discusses wild genetic resources and some of the associated issues. Next we discuss emerging institutional arrangements that provide for the evolution of well-defined property rights to wild genetic resources or alternatively to contracts that allow any resource rents to be captured. Third, the implications of property rights for the commercialization of genetic resources are addressed and some contracting problems raised. We observe that the assignment of property rights to wild genetic resources is as yet incomplete and significant transitional problems exist. We then examine the question of how important the incentives provided by property rights are likely to be as an incentive to mitigate overall biodiversity decline. The preliminary evidence suggests that the mitigating effect of the establishment of property rights to genetic resources may be quite small, due in part to the redundancy of species, as a species often is widely distributed. In this circumstance, the financial values of genetic resources are unlikely to be sufficient to protect biodiversity from further decline, since, as anecdotal evidence suggests, the use value of the marginal unit of biodiversity as an input in pharmaceuticals and related products is much smaller than the opportunity or protection costs of the lands on which biodiversity habitat resides.

Overview

It is widely agreed that wild genetic resources, the genetic constitutions of plants and animals, have substantial social and economic value as repositories of genetic information. Today, genetic information provides direct and indirect inputs into plant breeding programs, into the development of natural products including drugs and pharmaceuticals, and into increasingly sophisticated applications of biotechnology. Recognition of the potential of wild genetic resources in the development of drugs has, for example, led the United States National Cancer Institute to initiate a massive plant collection project, which is a follow-up to an earlier project in the 1970s seeking to identify drugs effective against a variety of cancers and also looking at the immune system effects of various drugs. Perhaps the most well-known examples of widely used drugs that have been developed from plants are two important anti-cancer drugs (vincristine and vinblastine) that are derived from the rosy periwinkle found in tropical Madagascar. In the 1980s the program was expanded to include a much wider set of screening activities to potentially identify compounds useful in a much broader set of applications. Other activities are also underway, the most well known of which is the agreement between the recently created Instituto Nacional de Biodiversidad (INBio) of Costa Rica and Merck Inc., a major pharmaceutical firm. This agreement is discussed further below.

With the recent development of major breakthroughs in biotechnology, the future potentiality for the development of useful drugs from wild plant constitutions seems even more promising. Species that have no current commercial application or useful natural chemicals, or are as yet undiscovered, nevertheless may have substantial value as a repository of genetic information that may someday have important commercial applications.

If preservation were costless, all genetic resources would be preserved. However, as the pressures on the habitat rise due to alternative uses, the costs of protection and preservation also rise. Until recently, preservation of genetic resources was essentially costless and all could be maintained. Two aproaches – *in situ* and *ex situ* – have been used to protect these acknowledged values. The *in situ* approach leaves the species in their natural habitats, whereas the *ex situ* approach involves permanent collections such as zoos, botanical gardens and the preservation of seeds and other genetic material in a controlled environment such as germplasm banks. Although the *ex situ* approach has the advantage of generally lower costs, this approach is feasible for only a small fraction of species. This approach, obviously, cannot be used for species as yet unknown (Harrington and Fisher, 1982). Furthermore, the *ex situ* approach preserves selected species, not ecosystems and thus risks the longer-term losses of species that are reliant upon the symbiotic relationships within the ecosystem.

Although the destruction of a unique genetic resource base can occur from the consumptive use of a particular plant itself, in practice a much more ominous threat comes from the process of land use change. Land use changes that destroy existing habitat and phenotypes (individual plants and animals) can inadvertently drive to extinction potentially valuable genotypes (the information embodied in the genetic constitutions of plant and animal species), many as yet undiscovered, that are endemic to certain ecological niches. Furthermore, individuals and countries engaged in development of the land resource, having no unique claim to the returns of the genetic information embodied in the wild plants, will tend to ignore the potential economic value of the existing habitat as a repository for potentially valuable genetic resources. The destruction of genetic resources becomes an unintended consequence, an externality, of habitat-destroying land use changes, and the costs of investing in protection and preservation can become substantial. By contrast, a company that utilizes a wild genetic resource as input into the development of a useful commercial product that is protected through a patent-like right often captures substantial financial returns.

One result of the lack of private or national property rights has been that, until recently, almost all efforts directed toward preservation and protection of genetic resources were altruistic and most proposals for preserving genetic resources have involved actions by governments and the international community to preserve habitat (OTA, 1987). The usual approach is for environmental groups and the governments of industrial countries to try to persuade Third World governments to protect habitats, such as tropical rain forests, that are rich in genetic resources. Although this approach has experienced some success, with the establishment of various parks and preserves, in many Third World countries there is an indifference to seriously protecting habitat preserves and the protection is often haphazard at best.

Toward the evolution of property rights

An important element to the notion of creating property rights to wild genetic resources is the concept that, while the phenotypes are entities that have rivalry and exclusivity in consumption, genotypes, the information embodied in the genetic constituents of the plant or animal, exhibit non-rivalry in consumption and require arrangements similar to those protecting intellectual property to provide exclusivity. As with intellectual property, exclusivity with an associated 'ownership' can, in principle, be achieved.

An earlier article (Sedjo, 1992) argued that two types of mechanism have been used to alleviate the inefficiency caused by an absence of property rights. One mechanism is the evolution of legally established property rights in response to changes in both the benefits associated with property rights and the costs of enforcing the rights as posited by Demsetz (1967). The second mechanism is suggested by the Coase Theorem (Coase, 1960), whereby an inefficient market outcome due to the existence of externalities may be 'corrected' through negotiation between the affected parties when transactions costs are not prohibitive.

Regarding the extension of property rights, the simple legal rule is that inventions are patentable, but discoveries are not. This distinction has always been present in United States case law, although the law does not explicitly treat 'discovery' as different from 'invention' (Beier *et al.*, 1985) and some legal decisions seem to deviate from this principle. By this criterion, wild genetic resources, being viewed as a 'discovery', would be less likely to receive patent protection than would engineered, modified, or improved genetic resources that might be viewed as an 'invention'. The establishment of exclusive rights through patents typically require the utilization of the wild genetic resource as an input to the development of an improved plant, or through the development of a unique (patentable) process for extraction or synthesis.

In order to maintain exclusivity in this legal environment, plant breeders have traditionally focused attention on hybrids, since the parental lines of the hybrid can be guarded and the seeds do not reproduce true to type. Similarly, it has been argued that American pharmaceutical producers have concentrated on processes and synthetics, in part because they had no unique rights to natural plants (Farnsworth, 1988).

However, with the advent of substantial improvements in biotechnology, property rights in the form of patent protection have tended to evolve to accommodate benefit-generating innovations. For example, a decision by an appeals board of the US Patent and Trademark Office has extended protection within the USA to allow patents for genetically engineered plants, seeds and tissue culture (Sun, 1985). The rationale is that, since it is now possible to describe a plant with the same precision as one can a machine, the difficulties of property rights assignment and enforcement are now dramatically reduced. It should be noted, however, that the courts have not all been consistent in their interpretations. For example, a recent decision extending property rights for genetic resources is that of a California Court of Appeals, which ruled that, if research reveals that a patient's tissues may yield products of commercial value, the donor has a right to compensation. Although this decision appeared to extend the concept of property rights to unimproved genetic resources under certain conditions, it was later reversed by the California Supreme Court.[1] Even

---

[1] Moore v. The Regents of the University of California, 249 Cal. Rptr. 494 (Cal. App2 Dist. 1988). This decision was overruled by the California Supreme Court in Moore v. The Regents of the University of California, 51 Cal. 3d 120, 792 p2 479, 15USPQ, 1753 (1990).

more recently controversy has raged over attempts of US companies to patent human gene sequences (Macilwain, 1993).

The recently concluded Biodiversity Convention (UNEP, 1992), which came out of the Rio Summit, can be viewed as an effort to expand the possibilities for providing patent-like rights for wild plants. The Convention appears to give standing to the concept that countries' sovereign rights include rights to control and compensation for genetic materials obtained from their territory. More generally, the Convention appears to extend some types of rights to indigenous genotypical knowledge, as well as for the genotype per se.

The second mechanism that could be utilized in confirming 'rights' to the genetic constituents of naturally occurring species is the Coasian contract. Under this view, contracts can internalize the external effects if transaction costs are sufficiently small. In earlier times plants and animals were simply taken by collectors. Within the last few years, however, contractual arrangements have begun to offer countries with rich endowments of natural genetic resources the potential for substantial compensation in return for access and the utilization of the genetic constituents of wild genetic resources found in these countries. These contractual arrangements require no new property rights, but rather they utilize the ordinary legal instrument of a contract. In effect, a source country trades the right to collection in return for a guarantee of some portions of the revenues generated by commercial development of a product that utilizes a genetic constituent from a wild genetic resource collected within the country.

Collection agreements have been evolving, presumably reflecting the perception of higher returns associated with capturing the externality. As noted above, the most well known of these types of arrangement is that between the Instituto Nacional de Biodiversidad (INBio) of Costa Rica and Merck Inc. Under this agreement Merck will provide INBio with an initial payment of $(US)1M over the next two years in return for the right-of-first refusal on large numbers of indigenous genetic resources. An additional royalty payment is due INBio for any commercial sales of products developed from the genetic materials.

Toward the commercialization of biodiversity: some contracting problems

Within the framework provided above, one focus has been on the characteristics of efficient contracts that provide incentives for conservation. An important consideration is to ensure that, from the conservation point of view, the contracting process provides appropriate incentives for the protection of habitat, e.g. the incentives at least include compensation for the opportunity costs of not undertaking habitat destructive activities. These should provide the land-owning entity with incentives to properly value the current and future expected returns to biodiversity as an input into pharmaceuticals in any land use decisions.

Relevant issues influencing the contract include perceptions, expectations and strategic considerations. As noted by Simpson and Sedjo (1993), these would include the relative risk aversion of the contracting parties, comparative expectations of future revenues, moral hazard considerations such as perceptions by sellers of the ability and willingness of buyers (product developers and marketers) to cheat on royalty payments, and concerns by buyers that desired plant sources be maintained and adverse selection considerations such as buyers' insistence that sellers certify the quality of the resource they offer. These considerations will influence the mix of payment arrangements. For example, the contractual form that appears to be emerging is that of a small up-front payment, with a royalty provision. Such an arrangement provides for the sharing of risk. In addition, it might be hypothesized, for example, that, although tropical countries might prefer up-front payments to longer-term royalties, the common arrangement of significant royalty arrangements reflects the more pessimistic assessment of the users of genetic resources of the expected returns than those of suppliers.

In addition, there is the issue of comparative advantage and the economics of vertical integration. The production process runs the spectrum from plant collecting, compound extraction, preliminary and secondary screening and assaying, fractionation or isolation, material development (and perhaps synthesis), patent applications and awards, pharmacological trials, clinical trials, regulatory approval and marketing and distribution. At what point in this complex process should the material seller transfer the functions to the developer?

Will property rights to genetic resources contribute to reduced biodiversity decline?

The previous sections have discussed ways in which property rights to the user-values of biodiversity may provide incentives for biodiversity habitat protection. In this section we examine whether the evidence supports the view that the establishment of property rights of the type discussed above will provide incentives to reduce biodiversity decline.

The evidence of the financial values of wild genetic resources is limited.

Collections sponsored by the US National Cancer Institute during the 1970s and 1980s provided only very nominal compensation to provider countries, although a fee was often provided to the collector or a local educational institute. However, if the payment simply covers the costs of collection, no resource 'rent' is generated. If no rent or surplus is realized, then the financial value of the genetic resource approximates to zero. Under these circumstances no financial incentive for protection is found, even in the presence of property rights.

Laird (1993) indicated that commercial collectors receive from $50 to $200 per sample. Aylward *et al.* (1993) in their case study of the Merck-INBio Agreement suggested that Merck's full payment was an average of between $333 and $1000 for each of the samples received. Payments of this magnitude may involve some 'rents' over and above collection; however, INBio also incurs costs for preliminary extraction, permanent inventorying and other services.

A necessary condition for property rights to provide incentives to protect wild genetic resources and therefore biodiversity generally is that ownership of those rights has financial value. Property rights to items without value provide no financial incentives. Thus, for property rights to wild genetic resources to provide an incentive to conserve requires that the wild genetic resources have financial value. If there is widespread biological redundancy in the system, as when the same wild genetic resource is widely distributed, the financial value of the marginal genetic resource of that type is likely to be negligible since it lacks uniqueness.

A general problem is that of determining when the anecdotal information about the prices of transacted wild genetic material is sufficiently high to indicate a rent. In the absence of appropriate property rights and proper contracts, prices would still represent only collection costs. This appears to be the current situation for much of the collection that is being undertaken. One problem may be the incomplete nature of the property rights institutions as they have emerged thus far. For example, within Costa Rica, INBio does not appear to be the sole collector and distributor of the wild genetic resources. In effect, the existing situation might be viewed as one in which there is still relatively open access to the resource. It appears that a number of separate independent agents are authorized to undertake collection. Under these circumstances resource rents could still be eroded to zero and financial incentives to protect habitat would be minimal. Another issue relates to the treatment of the potential royalty payment that would be received should the genetic material prove to be the source of an important drug. For a major drug the payment would be very substantial and could be viewed as being all economic rent.

In addition, if we view property rights 'emerging' to allow for the capture of rents, then in the early period after the emergence of appropriate property rights/contracts, the rental capture may be modest, even where the property rights and/or contracts are proper for capturing rents. As a practical matter the level of rent capture makes it even more difficult to determine whether the initial payments are modest. For example, modest *ex ante* payments, just covering collection and processing costs, would suggest little rent. However, should the actual royalty payments be large (on the average), an *ex post* assessment would determine the rents to have been large.

In summary, in order for rents to potentially exist, (a) access must be restricted (or clear property rights must exist) and (b) the material must be genuinely scarce. Overall, it appears that the evolving institutions have not yet eliminated the open-access nature of the wild genetic resource commons. For example, as noted, although the INBio–Merck agreement provides that INBio will provide only Merck with genetic materials (extracts), INBio does not itself have exclusive rights to collect and 'sell' Costa Rican wild biodiversity. Other collectors are allowed. Thus, access to the common Costa Rican biodiversity pool has been reduced but an element of its open-access nature exists. In these circumstances the market price of the genetic material transaction would not be expected to reflect its true scarcity rent, should any exist, but would probably be closer to the collection costs. This general problem appears to be exacerbated by the reality that the material is often *not* genuinely scarce. There is a redundancy of locations in which to find many of the biota. If the material can be found in several countries throughout Central America, then its very redundancy indicates that there is no scarcity rent to be collected. This circumstance reduces the price any pharmaceutical firm would need to pay. Nevertheless, the broad availability of the wild genetic resource also suggests its lack of endangerment.

Hence, we conclude the following. (a) A system for the assignment of property rights to wild genetic resources is as yet incomplete. Significant transitional problems exist. (b) Furthermore, current evidence indicates that the financial values of genetic resources could be quite low by virtue of their redundancy and thus limiting the potential of a property rights system to provide substantial conservation and protection incentives.

References

Aylward, B. A., Echeverria, J., Fendt, L. and Barbier, E. B. (1993). The economic value of species information and its role in biodiversity conservation: case studies of Costa Rica's National Biodiversity Institute and pharmaceutical prospecting. Unpublished report by the London Environmental Economics Centre and the Tropical Science Center in collaboration with the National Biodiversity Institute of Costa Rica, July.

Beier, F. K., Crespi, R. S. and Strass, J. (1985). *Biotechnology and Patent Protection: An International Review*. Organization for Economic Cooperation and Development, Paris.

Coase, R. (1960). The problem of social cost. *Journal of Law and Economics*, **4**, 1–44.

Demsetz, H. (1967). Toward a theory of property rights. *American Economic Review*, **57**, 347–73.

Farnsworth, N. R. (1988). Screening plants for new medicines. In E. O. Wilson, ed. *Biodiversity*, pp. 83–97. National Academy Press, Washington, DC.

Harrington, W. and Fisher, A. C. (1982). Endangered species. In P. R. Portney, ed. *Current Issues in Natural Resource Policy*, pp. 117–48. Johns Hopkins Press, Baltimore, MD, for Resources for the Future.

Laird, S. (1993). Contracts for biodiversity prospecting. In W. V. Reid, S. A. Laird, C. A. Meyer, R. Gamez, A. Sittenfeld, D. H. Janzen, M. A. Gollin and C. Juma, eds. *Biodiversity Prospecting*, pp. 99–139. WRI Publications, Washington, DC.

OTA (Office of Technology Assessment) (1987). *Technologies to Maintain Biological Diversity*. US Government Printing Office, Washington, DC.

Macilwain, C. (1993). OTA panel opens inquiry into patenting of genes. *Nature*, **362**, 386.

Sedjo, R. A. (1992). Property rights, genetic resources, and biotechnological change. *Journal of Law and Economics*, **35**, 199–213.

Simpson, R. D. and Sedjo, R. A. (1993). The commercialization of indigenous genetic resources: values, institutions, and instruments. Unpublished paper delivered to conference on Market Approaches to Environmental Protection, Stanford University, 4 December.

Sun, M. (1985). Plants can be patented now. *Science*, **230**, 303.

UNEP (United Nations Environmental Programme) (1992). Convention on Biological Diversity.

Part C

Diversity decline as policy failure

8

Economic progress and habitat conservation in Latin America

DOUGLAS SOUTHGATE

Introduction

Natural habitats in Latin America, which harbor a large portion of the world's flora and fauna, are being uprooted at an alarming pace. More than a quarter of the globe's closed forests are in the region and numbers of rare and threatened plant and animal species are huge (WRI, 1992, pp. 286–7 and 304–7). Especially outside the Amazon Basin, rates of agricultural land clearing are high relative to remaining tree-covered area. Myers (1988) has identified ten hotspots in which the threat to biodiverse tropical forests is particularly severe. Four of those zones, including two of the three 'hottest' ones, are in the western hemisphere.

In part, destruction of rain forests and other ecosystems is the result of market failure. Many of the services provided by those resources (e.g. climate regulation and the harboring of flora and fauna) share the two principal characteristics of public goods (Samuelson, 1954). First, consumption is non-rivalrous. This means that each and every consumer would have to pay the maximum amount he or she attaches to a small increment in resource quality or availability in order for private ecosystem management to be efficient. Second, markets for ecosystem services are poorly developed because costs of excluding access are high.

Measures can be taken to deal with misallocations caused by non-rivalry in consumption and non-exclusivity. Perfect price discrimination is unlikely to occur among those who benefit from biodiversity conservation and climatic stability. Nevertheless, funds raised from around the world ought to be sufficient to save many wild habitats, by paying off agricultural colonists and other agents of ecosystem destruction. Moreover, markets for selected environmental services are beginning to emerge. For example, agreements to raise trees for the purpose of carbon sequestration are being worked out between the owners of deforested land that yields little or no

agricultural income and public utilities and other energy-intensive industries located in countries that tax carbon emissions.

The distorted private returns to agricultural land clearing

The performance of whatever green markets emerge, and indeed the fate of
biodiverse ecosystems in Latin America, will hinge on the private net
benefits associated with tropical deforestation. If they are large, payments
to arrest agricultural land clearing will have to be correspondingly large. If
small, payments may be modest.

Obviously, private net benefits rarely reflect environmental costs.
Instead, they comprise the difference between agricultural and forestry
rents, evaluated at prevailing market prices. Furthermore, they have been
inflated over the years because public policies have distorted the relative
returns to forestry and agriculture.

Forestry in Latin America has suffered from the policy regime applied to
encourage economic development through import substitution and industrialization. Exports of unprocessed timber have been taxed and restricted
in order to keep raw material costs low for wood products industries, which
have benefited in addition from tariffs and restrictions on imports of
plywood and other finished products. The practical effect of log export
bans, which many countries have applied, has been to create or to reinforce
oligopsony in local timber markets. As a result, domestic timber prices have
been held to a fraction of international values in many places and
landowners' incentives to plant trees and to manage forests have been
weakened.

To be sure, growth in other parts of the rural economy, including
agriculture, has been held back by macroeconomic and sectoral policies
that favored urban-based industrial development by keeping commodity
prices artificially low. Currency over-valuation, which was a general feature
of economic life in the region until very recently (Krueger *et al.*, 1988), has
discouraged commodity exports and made it difficult for domestic producers to compete with imports. Likewise, food price controls have had
direct and negative impacts on crop and livestock production.

At the same time, though, policies biased toward agriculture have been
applied in frontier areas. Directly or indirectly, the Brazilian government
has subsidized livestock production in the Amazon, which has stimulated
limited deforestation in some areas (Mahar, 1989). Throughout the hemisphere, an old legal tradition of granting tenure to individuals who make
'productive use' of tree-covered land (by clearing it for agriculture)

continues to live on, at least informally. Where this rule is in force, the returns to agriculture include a diminished risk that one's property will be invaded and expropriated.

Policy reform is leading to a more level playing field for forestry and agriculture in those parts of Latin America traversed by agricultural frontiers. Agrarian reform agencies, which have organized colonization of the region's tropical forests since the 1960s, are doing away with agricultural land clearing as a prerequisite for formal tenure. Likewise, Brazil's deforestation subsidies are largely a thing of the past.

Progress toward elimination of log export bans and other policies that depress timber values has been slow in many countries. Unless prices are allowed to approach international levels, however, nobody will perceive that deforestation carries an opportunity cost or that spending money on the management of timber resources is warranted.

The importance of agricultural intensification

Another way to reduce the private net benefits of deforestation, and thus to protect the natural ecosystems found beyond agriculture's extensive margin, is to intensify crop and livestock production. As yields increase, commodity prices decline, all else remaining the same. Farmers and ranchers located far from markets observe that the small margin between their revenues and production costs shrinks dramatically, often disappearing entirely or turning negative. When this occurs, there is little or no incentive to create new agricultural land and the extensive frontier can actually recede.

As recent Chilean experience demonstrates, agricultural intensification can be achieved through increased mechanization, irrigation, and the use of chemical fertilizers and also through the dissemination of improved varieties and cultivars (Arensberg *et al.*, 1989). Where all this is done, agriculture's extensive margin remains relatively stable. During the 1980s, for example, commodity demand rose dramatically in Chile, due to population growth averaging 1.7% a year and a boom in agricultural exports. Nevertheless, there was practically no geographical expansion in the country's agricultural economy because yields were rising fast (Southgate, 1994).

As a simple regression analysis demonstrates, the inverse relationship between agricultural intensification and outward shifts in the agricultural frontier holds for Latin America as a whole. Annual percentage growth in cropland and pasture during the middle 1980s (AGLNDGRO) is the

D. Southgate

regression's dependent variable and the right-hand side variables include annual percentage growth in population, exports, and yields (POPGRO, EXPGRO, and YLDGRO, respectively) as well as a dummy factor indicating that land not yet cleared is poorly suited to agriculture (NOLAND).

With data for 23 countries, ordinary-least-squares estimation of the regression model (Southgate, 1994) yields the following results:

$$\text{AGLNDGRO} = 0.463 + 0.249 \text{ POPGRO} + 0.031 \text{ EXPGRO}$$
$$(0.161) \quad (0.066) \qquad\quad (0.014)$$
$$(2.876) \quad (3.773) \qquad\quad (2.214)$$
$$- 0.198 \text{ YLDGRO} - 0.641 \text{ NOLAND};$$
$$(0.033) \qquad\quad (0.205)$$
$$(- 6.000) \qquad\quad (- 3.127)$$

$$\text{adjusted } r^2 = 0.669, \qquad F = 12.098.$$

The signs of all parameter estimates are consistent with what should be expected and interpretation of the coefficients is straightforward. In a country where natural conditions do not favor frontier expansion (i.e. where the value of NOLAND is 1 instead of 0), the annual increase in cropland and pasture is expected to be 0.641 percentage points lower than would be the case if soils that lend themselves to crop or livestock production were unoccupied. Furthermore, a $Z\%$ increase in yields either offsets nearly four-fifths of the agricultural land clearing induced by $Z\%$ population growth or compensates for $6 \times Z\%$ growth in agricultural exports.

Commodity demand in Latin America is bound to increase since population growth is continuing (notwithstanding recent declines in human fertility), agricultural exports are increasing (as discriminatory policy regimes are reformed), and incomes are beginning to rise. If crop and livestock yields do not grow rapidly enough, future agricultural land clearing is inevitable.

The fiscal, human, and environmental opportunity costs of misguided economic policies

As the pursuit of economic development strategies predicated on import substitution and industrialization exemplifies, Latin American governments have been quick to interfere with market forces. The sacrifice in GDP alone, caused by distorted price signals, has been staggering. In addition, economic progress has been held back because financial resources, which could have been used to make fundamental improvements in productive capacity, were instead dissipated in the form of subsidies and other governmental interventions in the market place.

This later consequence is illustrated by public expenditures on agriculture in Ecuador. Millions of dollars continue to be spent each year on irrigation subsidies, which benefit a small and relatively affluent minority of the country's farmers, and plans to spend up to $(US)2 billion on inefficient new projects, which are supported by individuals eager to receive subsidized water, are being pursued. Meanwhile, governmental support for technology development and dissemination, which are essential for environmentally sound agricultural intensification, has dwindled steadily. Having declined by 7.3% a year from 1975 through 1988, real spending on agricultural research in Ecuador is now less than 0.2% of agricultural GDP, which is low by the standards of neighboring countries (Southgate and Whitaker, 1994). It is hardly surprising that agricultural yields have been low and stagnant and that increases in commodity demand, associated with population and export growth, have induced high rates of agricultural land clearing (Southgate, 1994).

A closely related fiscal consequence of misguided economic policies has been inadequate formation of human capital. Since budgets have been drained to pay for subsidies (and also for bureaucratic bloat and excessive military spending), less money has been left over for education, public health services, and so forth. Support for human capital formation in rural areas has been especially deficient.

This has had direct impacts on the welfare of the rural poor and also on the environment. There are millions of people in the Latin American countryside who have found themselves poorly equipped in recent years to compete in an economy undergoing modernization. All too often, their choices have been reduced to two options. The first is to crowd into the slums that ring the region's major cities, hoping to find a low-paying job. The second is to subsist on a small farm established on a barren hillside or carved out of a tropical forest. Neither option is remunerative, and choosing the former adds to the deterioration of urban life whereas the latter results in the destruction of land, tropical forests, and other renewable resources.

False starts

Alarmed by biodiversity loss in Latin America, national governments, donor agencies, and environmental organizations are seeking to protect the region's natural ecosystems. In recent years, support has grown especially strong for Integrated Conservation and Development Projects (ICDPs), in which park protection is coupled with the promotion of environmentally benign activities (e.g. agroforestry, ecotourism, and extraction of non-timber products) in so-called buffer zones around officially designated nature reserves. For example, spending by the US Agency for International Development (AID) on ICDPs and closely related environmental initiatives in the western hemisphere grew by nearly 50%, in real terms, between fiscal years 1990 and 1992.

ICDPs have been plagued by design and implementation difficulties. As a thorough review recently carried out for the World Bank, AID, and the World Wildlife Fund shows, the goal of habitat protection has been difficult to reconcile with local communities' aspirations for a better life in Africa, Asia and Latin America (Wells and Brandon, 1992).

But the most important shortcoming of a conservation strategy in which ICDPs are a centerpiece is that linkages between poor economic performance and renewable resource depletion are not addressed. As backing for habitat protection has increased, funding for education, research, and extension in rural areas has fallen. For example, AID's agricultural programs in the western hemisphere have been declining for several years. Likewise, real spending by the Consultative Group on International Agricultural Research fell by 35% and the system's senior staff shrank by more than a fifth between 1988 and 1993. During the same period, budgets and staffing levels for the International Center for Tropical Agriculture (CIAT), the International Maize and Wheat Improvement Center (CIM-MYT), and the International Potato Center (CIP), which all have head-quarters in Latin America, went through similar declines (CGIAR, various years).

A definitive judgement cannot be made on the effectiveness of conservation strategies predicated on ICDPs. This is simply because no such project has ever been implemented successfully in any part of Latin America where the threat of encroachment on natural ecosystems is imminent. Even without the results of a true test at hand, though, one expects that attempts to protect parks and to keep land use in surrounding buffer zones environmentally stable will look puny once the onslaught released by inadequate formation of human capital and low support for agricultural research and extension arrives on the scene.

A strategy for economic development and habitat conservation

As it was developed and as it continues to be practiced in North America, northern Europe, and other affluent parts of the world, environmental economics concentrates on the market failure induced by non-rivalry in consumption and open access to natural resources. Since most environmental economists work in rich countries (where, by definition, human capital is abundant, the economy's science and technology base is strong, and infrastructure is at an advanced stage of development), relatively little has been written about the natural resource issues that arise in poor areas, which are distinguished primarily by the scarcity of wealth that can be substituted for natural resources.

In many respects, the prescriptions for dealing with market failure, which can be found in any environmental economics textbook, have universal validity. However, as a number of economists emphasize, habitat destruction, specifically, and renewable resource depletion, generally, sometimes have little to do with market failure. Instead, misguided governmental interference with private property rights and market forces is often the major culprit (Panayotou, 1992; Repetto and Gillis, 1988; Southgate and Whitaker, 1994). Rather than coming up with effective interventions to deal with externality problems, which can be a vexing task, governments often can contribute most to the wise use and management of natural resources by reforming the policies that diminish incentives for conservation.

In this chapter, special emphasis has been placed on the importance of human capital and other non-environmental assets. The argument has been made that, because of the budgetary drain inherent in misguided economic policies, too little has been done to strengthen the scientific base underpinning crop and livestock production, and support for education, especially in the countryside, has been inadequate. One result of this misdirection of resources has been to make geographical expansion the agricultural economy's primary response to increases in commodity demand, induced by population, income, and export growth. Another consequence has been to add to the number of rural poor who must colonize tropical forests and other natural ecosystems in order to survive.

There is a strong need for empirical research on the sensitivity of deforestation rates to the opportunity cost of the effort that agricultural colonists dedicate to clearing land and raising crops and livestock. Such research will probably show that raising the earning power of the rural poor could help to save a great deal of Latin America's biodiversity. Studies carried out around the world indicate that the returns to agricultural colonization are marginal, at best. If colonists had the training and education required to work elsewhere for more pay, many of them would choose to leave natural habitats rich in flora and fauna alone.

In this sense, there is a fundamental complementary between economic progress and biodiversity conservation in the western hemisphere.

D. Southgate

References

Arensberg, W., Higgins M., Asenjo, R., Ortiz, F. and Clark, H. (1989). Environment and natural resources strategies in Chile. US Agency for International Development, Santiago.

CGIAR (Consultative Group on International Agricultural Research) (various years). *Financial Reports*. Washington, DC.

Krueger, A., Schiff, M. and Valdés, A. (1988). Agricultural incentives in developing countries: measuring the effect of sectoral and economy-wide policies. *World Bank Economic Review*, **2**, 255–71.

Mahar, D. (1989). *Government Policies and Deforestation in Brazil's Amazon Region*. International Bank for Reconstruction and Development, Washington, DC.

Myers, N. (1988). Hotspots in tropical forests. *Environmentalist*, **9**, 1–20.

Panayotou, T. (1992). *Green Markets: The Economics of Sustainable Development*. Institute of Contemporary Studies Press, San Francisco.

Repetto, R. and Gillis, M. (1988). *Public Policies and the Misuse of Forest Resources*. Cambridge University Press, Cambridge.

Samuelson, P. (1954). The pure theory of public expenditure. *Review of Economics and Statistics*, **36**, 387–9.

Southgate, D. (1994). Tropical deforestation and agricultural development in Latin America. In K. Brown and D. Pearce, eds., *The Causes of Tropical Deforestation: The Economic and Statistical Analysis of Factors Giving Rise to the Loss of Tropical Forests*, pp. 134–44. University College London Press, London; University of British Columbia Press, Vancouver.

Southgate, D. and Whitaker, M. (1994). *Economic Progress and the Environment: One Developing Country's Policy Crisis*. Oxford University Press, New York.

Wells, M. and Brandon, K. (1992). *People and Parks: Linking Protected Area Management with Local Communities*. International Bank for Reconstruction and Development, Washington, DC.

WRI (World Resources Institute) (1992). *World Resources, 1992–1993*. Oxford University Press, New York.

9

Prudence and profligacy: a human ecological perspective

MADHAV GADGIL

People and resources

People are evidently responsible for the current pace of loss of biodiversity. The search for the root causes of this loss may then profitably begin with an enquiry into how people relate to their base of natural resources. Dasmann (1988) provided an insightful analysis of the different ways in which they do so. He identified people at the two extremes of what is undoubtedly a continuum as ecosystem people and biosphere people. Ecosystem people depend largely on their own muscle power, and that of their livestock, to gather, produce and process most of the resources they consume. The bulk of these therefore come from a limited resource catchment; from an area of around 50 km^2 around their homesteads. Moreover ecosystem people have characteristically used their resource catchments over long periods, often several generations. Many tribal, peasant, pastoral, rural artisanal communities of India fit this description. Biosphere people, conversely, have extensive access to additional sources of energy such as fossil fuels or hydroelectricity. This enables them to transport and transform large quantities of material resources from all over the world for their own use. Their resource catchments are vast. They are continually tapping newer and newer resources, for instance nuclear power, from newer localities, e.g. drilling for oil in the deeper seas. They are very much part of an increasingly globalising market economy. Most citizens of the First World, and the Third World elite behave as biosphere people.

To these two categories of Dasmann, one may add yet another, that of ecological refugees. Ecological refugees are ecosystem people, deprived of access to their traditional resource base, who are forced to colonise new localities, where they continue to depend largely on human and animal muscle power to gather, produce and process resources (Figure 9.1). Their resource catchments remain limited, but these are no longer ecosystems with which they have been integrated over generations. They therefore do

M. Gadgil

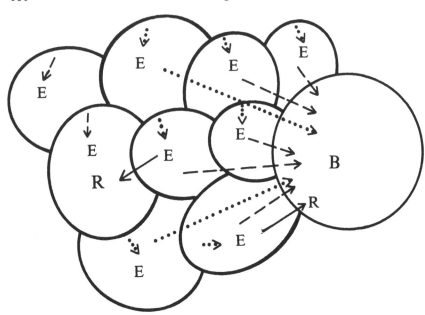

Figure 9.1. Schematic representation of fluxes of people and resources. Biosphere people (B) organise resource fluxes from over a wide resource catchment. Ecosystem people (E) ultilise resources from more restricted catchments. Strong (——>) or weak (···>) resource fluxes directed toward concentrations of biosphere people lead to resource depletions that convert ecosystems people into ecological refugees (R), who either move (→) toward areas from which biosphere people have not as yet drawn much resources as forest encroaches or toward concentrations of biosphere people as urban squatters.

not have the attachments, nor the knowledge, nor the motivation to use the resources of these new catchments in a prudent fashion. White colonisers of North America were largely such ecological refugees of the resource crunch that afflicted Europe during the Little Ice Age (Crosby, 1986). So must have been the people who gradually colonised whole series of Pacific islands over the last millennium (Diamond, 1991). Examples of present day ecological refugees include peasants moving into the rain forest of the Amazon basin in Brazil, or Western Ghats in India. Another wave of ecological refugees today creates the shanty towns of Third World cities, or lives as illegal, immigrant labour working on farms in the United States.

What promotes prudence?

People, by and large, pursue self-interest. They are unlikely to be motivated to use a resource base in a prudent, sustainable fashion: (a) if their resource

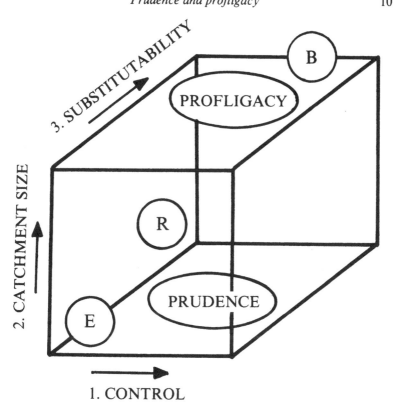

Figure 9.2. The three major factors governing motivation towards prudent or profligate resource use are (1) extent of control over the resource base, (2) the size of catchment from over which resources are derived, and (3) the extent of substitutability of an exhausted resource element by another. Currently none of the three components of the human populations, the ecosystem people (E), the biosphere people (B) or ecological refugees (R) are so situated in this three dimensional space as to be motivated to practice prudence.

catchments are vast, so that degradation of any particular part of the catchment affects them very little; (b) if they have open before them possibilities of substitution as any one resource element is depleted; or (c) if their control over the resource base is tenuous, so that others may at any time deplete a resource they value, even if they used it in a restrained fashion. Indeed, I believe that profligacy is likely when any one of these three conditions obtains. It is only when people perceive their resource catchments as limited, the possibilities of substitution of an exhausted resource element as not readily accessible, and their own control over the resource base as extensive, that they will be motivated to use the resource base in a prudent fashion (Figure 9.2).

Ecosystem people, rooted in a locality, and retaining control over their resource base are most likely to fulfil all three prerequisites for sustainable resource use; therefore they tend to behave in ways conducive to the conservation of the biodiversity of their own localities. Today, hardly any ecosystem people retain control over resources; that has been or is being usurped by the more powerful biosphere people. But in the preindustrial world, many communities of ecosystem people are likely to have fulfilled all three conditions promoting prudence. It must be emphasised that when newly colonising a locality, people with hunting–gathering–fishing–subsistence agriculture technologies are unlikely to have fulfilled these conditions. At the frontier, they are likely to have perceived potential resource catchments as extensive, and faced with abundant resources perceived many possibilities of substitution. Such people would not be ecosystem people rooted in a locality in our terminology; they may be thought of as ecological refugees unlikely to behave prudently. The many examples of extinctions of species on islands such as Madagascar or in Polynesia are probably cases of profligate resource use by early colonising communities (Diamond, 1991).

Cultural traditions

All human communities would at some point in their history have been such colonisers. Initially they may have neither the motivation nor the knowledge of the resource base to arrive at regimes of sustainable use. In this period they may be responsible for, and witness, the elimination of many biological populations. If they do become rooted in a locality, and come to control its resource base effectively, they are likely to see themselves as being affected by resource overuse and gradually become motivated to use the resources in a prudent fashion. When so motivated they may develop some simple rules of thumb to promote conservative use through a process of trial and error. Joshi and Gadgil (1991) suggested that such a process of trial and error may be based on comparing levels of harvesting effort and yields in the recent past. This could lead to a decision rule such as the following.

1. Enhance harvesting effort: if an enhancement of harvesting effort in the past was accompanied by increased harvests (or in some other way left one better off); or if a reduction in the harvesting effort in the past was accompanied by reduced harvests (or in some other way left one worse off).

2. Step down harvesting effort: if an enhancement of harvesting effort in the past had left one worse off; or a reduction in harvesting effort had led to one being better off.

Such a decision rule, Joshi and Gadgil (1991) showed, can lead to a sustainable harvest, especially when the reduction in the harvesting effort takes the form of total protection of some resource element. Such protection may be afforded by creating refugia such as sacred groves or sacred ponds from which no harvests are made; it may involve total protection to all individuals of some taxa such as the keystone tree genus *Ficus*, widely protected in the Old World; or total protection to especially vulnerable life history stages such as colonial breeders at a heronary. Such measures evidently tend to promote conservation of biodiversity coupled to its sustainable use. Ecosystems people in many parts of the world indeed possess a variety of such cultural traditions of conservation practices (Gadgil and Berkes, 1991; Gadgil *et al.*, 1993). Indeed, even Polynesian islanders seem to have developed a variety of such prudent practices, following the initial spate of exterminations (Ruddle and Johannes, 1985). So also Madagascarans came to protect the lemur species as sacred animals after having been earlier responsible for the extinction of giant lemurs (Diamond, 1991; Jolly, 1980).

In traditional societies such practices were institutionalised through prescriptions based on religious beliefs. But that they may have been grounded in secular interests is brought out by the experience in northeastern hill states of India in recent years. The hunter–gatherer–shifting cultivator communities of these states at the junction of India, China and Myanmar exhibited a number of traditional conservation practices. These included protection of sacred groves, which might have covered as much as 10–30% of the land area, as well as practices of regulated harvests from other woodlands. On conversion to Christianity, largely in the 1950s, most of these communities abandoned the protection of sacred groves. But soon they discovered that whole-sale clear-felling had many adverse consequences, including increased fire hazard for settlement areas during the slash-and-burn operations. So many communities of the northeast, now converted to Christianity, have revived the protection of sacred groves, especially those encircling their settlements. In the new idiom they term these as 'safety' forests in place of 'sacred' forests; but the punishment for violating the taboos against cutting in the safety forest is identical with that in the pre-Christian era (Malhotra, 1990; N. S. Heman and M. Gadgil, personal observation). This rapid revival of traditions of conservation through a new idiom has occurred most noticeably in the northeastern parts of India where the local people retain a larger measure of control over their resource base, but it has also been taking place in many other parts of India (Malhotra and Poffenberger, 1989).

Domain of concern

Humans remained of necessity largely rooted to their own localities so long as all they consumed was gathered with their own labour, with scarcely any margin for usurpation of surplus by others. Husbanding of plants and animals began to change all that; it was now possible for surplus grain or meat on hooves to be transported over distances for use by people who had played no role in its production. Initially only a small elite of priests–warriors–bureaucrats formed the nucleus of this new class of biosphere people. Over time, this class has expanded as technological advances have permitted larger and larger volumes of resources to be transported over greater and greater distances. The possibility of bringing in resources from a distance simultaneously makes available the option of not utilising resources close to home. It has been suggested that humans have an inherent love of certain forms of natural landscapes (topophilia) and of a variety of living organisms (biophilia) (Wilson, 1984; Kellert and Wilson, 1993). Biosphere people then potentially have the choice of maintaining biologically diverse communities around themselves by lightly exploiting their immediate surroundings, while transferring the pressure of resource exploitation to more distant tracts. Indeed they have often engaged in such geographical discounting, since the level of any people's concern for the health of the ecosystem appears to fall off, often quite rapidly with distance from their habitation (Figure 9.3). Thus, Kautilya's *Arthashastra*, the 2000 year old Indian manual of statescraft, prescribes the maintenance of princely hunting preserves just outside the capital (Kangle, 1969), and Japanese today maintain 60% of their country under tree cover while importing timber from Southeast Asia and the Americas. Similarly, there have been attempts in the United States to export toxic wastes to countries such as Haiti. Very generally, the biosphere people have always attempted to shift the negative consequences of their resource consumption to the hinterlands, as far removed in space, and perhaps on other criteria such as ethnic and cultural affinity, as technology would permit.

Sequential exploitation

I would like to suggest that it is this disjunction between the domain of concern and the catchment for resources by biosphere people that is an

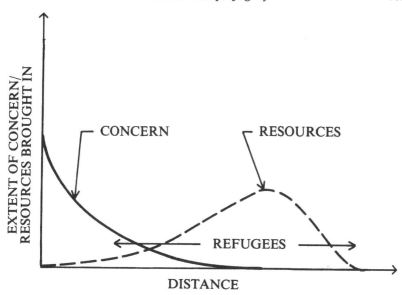

Figure 9.3. Biosphere people are concerned primarily with the health of ecosystems of localities close to their habitations, whereas they concentrate resource harvest from areas beyond the domain of concern. The resulting resource depletion in these distant areas creates ecological refugees who either move on to the periphery of areas of concentration of biosphere people, or into areas that have as yet been much less affected by resource extraction pressures by the biosphere people.

important root cause of the ongoing erosion of biodiversity in the present day world. Consider as an example of how this process operates the impact of the plywood industry on the tropical humid forests of India. The Indian plywood industry has received its raw material from the state-controlled forest lands at highly subsidised rates. Thus, mangoes growing wild in the forests of the Western Ghats are a prized source of fruit used in many condiments. But whole huge mango trees were conceded to the plywood industry at rates less than the value of a single year's yield of fruit. Economic theory tells us that resources for which rates of annual increment are low and profits to be made from harvest are high are susceptible to overexploitation. It is suggested that the state should therefore step in to appropriately tax resource use so as to bring down the demand (Dasgupta, 1982). Quite contrary to this prescription, India's state machinery has stepped in to further reduce resource costs and thereby encourage over-exploitation (Gadgil, 1992).

In this setting the plywood industry has tended at any time to concentrate its exploitation pressure on the resource element that at that time generates the highest levels of profits. Thus, each mill initially prefers to focus on the closest localities, on the most accessible terrain, on trees of largest girth, on the species that are most suitable for manufacture of plywood. As the largest trees of most suitable species from closest, most accessible localities are exhausted, the industry switches in a sequence to smaller trees of less suitable species from less accessible localities further away (Gadgil, 1991). In the process there has been a substantial depletion of biodiversity from over much of the tropical humid forest tracts not only of the Indian subcontinent but of the Andaman and Nicobar islands as well.

Ecological refugees

This process of sequential exploitation of biological resources to fulfil the needs of biosphere people that are growing without any limit largely affects terrain outside their domain of concern. Thus, the localities favoured by the rich and powerful in Indian cities are amongst the greenest areas of the country; and Germany maintains its Black Forest even as it draws on the humid forest of tropical Africa. Most affected in this process are the ecosystem people of the Third World countryside. Bamboo is a mainstay of a large number of India's rural artisans. It is also a raw material much favoured by the paper industry. While the industry has been receiving bamboo supplies essentially free of charge from state forests, its market prices have soared even as its availability has sharply declined due to industrial overexploitation. In the process many rural artisans have been badly affected and lost their only source of livelihood. Similarly, large numbers of peasants and tribal peoples have been displaced by river valley projects, without adequate compensation or rehabilitation, even as the beneficiaries from these projects have been receiving water and power at highly subsidised rates (Centre for Science and Environment, 1985). So, deprived of their traditional subsistence base, ecosystem people of the world are increasingly turning into ecological refugees, ending up either encroaching on forest lands or as squatters in urban shanty towns (Figures 9.1 and 9.3).

The ecological refugees have neither the motivation nor the knowledge to deal prudently with the new environments they are forced into. Moreover, the legal framework within which they operate often compels them to act in a destructive fashion. Thus, in India, forest land is state property, and is land from which people derive little support for their own livelihoods, since it is primarily devoted to highly subsidised supplies of resources to the commercial sector. Nevertheless, they can claim title to a deforested piece of land brought under cultivation that will contribute towards their subsistence. So, all over the country, ecological refugees are fragmenting and encroaching on natural, diverse rain forest communities to convert them into much poorer communities of husbanded plants and animals and their weedy associates. But these are people with no other options.

Such processes are taking place all over the biodiversity-rich countryside of the Third World. Resource demands of biosphere people – most citizens of the First World and the Third World elite – are eroding the resource base of ecosystem people of the Third World countryside. These ecosystem people, turned into ecological refugees, are in turn further exerting degradative pressures on the Third World countryside, as well as on its urban centres. The process has its roots in the ever-growing resource demands of the biosphere people, and their willingness to permit resource degradation in tracts outside their domain of concern. It is a process that began 10 000 years ago when the social division of labour first brought biosphere people onto the world scene. It has started gathering pace in the last three centuries of rapid technological progress, greatly enhancing the capabilities of biosphere people to usurp the resource base of the world's ecosystem people. With these same technological advances triggering off a population explosion in the Third World countryside, the rapidly growing numbers of ecological refugees are adding to the forces of biodiversity erosion (Figure 9.4).

Challenges before us

The response of the biosphere people to this threat to biodiversity has been to promote establishment of nature reserves in the Third World countryside, and to accumulate the genetic diversity of husbanded plants in *ex situ* storages. This is not a viable strategy, for the nature reserves so set up are increasingly surrounded by degraded ecosystems in which large numbers of ecological refugees are attempting to eke out a precarious subsistence (Figure 9.5).

In its stead we must try to put in place a strategy that would arrest the processes that are swelling the ranks of the world's ecological refugees. This calls for a moderation of resource demands of the biosphere people and willingness on their part to pay a far better price for the resources that are being extracted at the cost of the ecosystem people. There must be serious efforts to rehabilitate ecological refugees, to help them strike roots and revert to the status of ecosystem people. As Figure 9.2 implies, it is the ecosystem people who are most likely to bring practices of prudence to prevail, provided that they are given adequate powers to control their

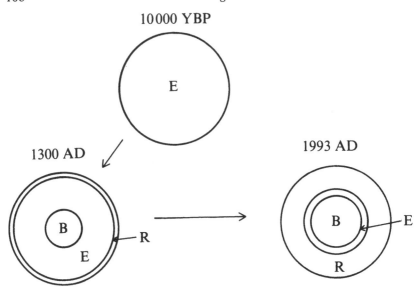

Figure 9.4. Historical changes in the relative proportions of ecosystem people (E), biosphere people (B) and ecological refugees (R). Prior to the beginnings of agriculture, all of humanity behaved as ecosystem people; with agrarian societies developed biosphere people who turned a small proportion of ecosystem people into ecological refugees. Today a further expansion in numbers and demands of biosphere people is leading to a rapid shrinkage of space for ecosystem people who are being increasingly converted into ecological refugees. YBP, years before present.

resource bases. Serious attempts must therefore be made to empower these people to take good care of the health of their ecosystems.

But a world with a permanent gulf between the biosphere people and the ecosystem people is not a sustainable world. The ecosystem people everywhere are striving to move into the ranks of biosphere people, even as they are instead being forced into the status of ecological refugees. In the long run then, a socially sustainable world would only be an equitable community of biosphere people. For such a community to be ecologically sustainable, biosphere people will have to strike roots all over the world and take good care each of their own bioregions.

References

Centre for Science and Environment (1985). *The state of India's Environment 1984–85. A second Citizen's Report.* Centre for Science and Environment. New Delhi.

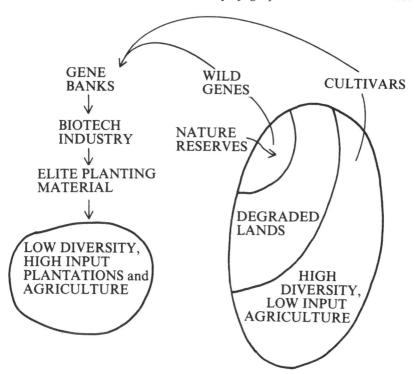

Figure 9.5. The current strategy of biodiversity management of biosphere people is to draw on the genetic resources of (a) cultivars in subsistence agriculture and (b) natural biota in islands of nature reserves in the midst of increasingly degrading Third World countryside. These genetic resources maintained in *ex situ* gene banks, serve to create elite planting material grown in intensively managed monocultures to yield high levels of profit.

Crosby, A. (1986). *Ecological Imperialism: The Biological Expansion of Europe.* Cambridge University Press, Cambridge.

Dasgupta, P. (1982). *The Control over Resources.* Oxford University Press, New Delhi.

Dasmann, R. F. (1988). Towards a biosphere consciousness. In D. Worster, ed. *The Ends of the Earth: Perspective on Modern Environmental History*, pp. 177–88. Cambridge University Press, Cambridge.

Diamond, J. (1991). *The Rise and Fall of the Third Chimpanzee.* Vintage, London.

Gadgil, M. (1991). Restoring India's forest cover: the human context. *Nature and Resources*, **27**(2), 12–20.

Gadgil, M. (1992). State subsidies and forest use in a dual society. In A. Agarwal, ed. *The Price of Forests*, pp. 132–7. Centre for Science and Environment, New Delhi.

Gadgil, M. and Berkes, F. (1991). Traditional resource management systems. *Resource Management and Optimization*, **18**, 127–41.

Gadgil, M., Berkes, F. and Folke, C. (1993). Indigenous knowledge for biodiversity conservation. *Ambio*, **22**, 151–6.

Jolly, A. (1980). *A World Like Our Own. Man and Nature in Madagascar*. Yale University Press, New Haven, CN, and London.

Joshi, N. V. and Gadgil, M. (1991). On the role of refugia in promoting prudent use of biological resources. *Theoretical Population Biology*, **40**, 211–29.

Kangle, R. P. (1969). *Arthashastra*. University of Bombay, Bombay.

Kellert, S. and Wilson, E. O. (eds.) (1993). *The Biophilia Hypothesis*. Island Press, Washington, DC.

Malhotra, K. C. (1990). Village supply and safety forest in Mizoram: a traditional practice of protecting ecosystems. In *Abstracts of the 5th International Congress of Ecology*, Yokohama, Japan, p. 439.

Malhotra, K. C. and Poffenberger, P. (eds.) (1989). *Forest Regeneration Through Community Protection: The West Bengal Experience*. Proceedings of the Working Group Meeting on Forest Protection Committees, Calcutta, 21–22 June, West Bengal Forest Department, Calcutta.

Ruddle, K. and Johannes, R. E. (eds.) (1985). *The Traditional Knowledge and Management of Coastal Systems*. UNESCO, Jakarta Pusat.

Wilson, E. O. (1984). *Biophilia*. Harvard University Press, Cambridge, MA.

10

Tropical deforestation: population, poverty and biodiversity

NORMAN MYERS

Introduction

Tropical moist forests, hereinafter referred to simply as tropical forests, contain at least 50% and possibly a still larger proportion of earth's species in only 6% of earth's land surface. The forests share another unique characteristic: they are being depleted faster than any other large-scale biome. So this is where the impending mass extinction will mainly occur – or be contained. If the forests are virtually eliminated within the next few decades, there will be a huge loss of species, regardless of our conservation efforts in the rest of the world; but if we manage to save most of the forests while allowing other major environments to be widely depleted, we shall largely avoid a mass extinction. So tropical forests are exceptionally critical to the biodiversity prospect writ large.

A single agent of deforestation, slash-and-burn cultivators, now destroys almost 100 000 km² of tropical forest each year, out of a deforestation total of roughly 150 000 km² (FAO, 1992; Myers, 1989, 1994). Not only do slash-and-burn cultivators cause the preponderant share of annual deforestation, but the proportion is growing. Thus, they are the prime source of the meta-problem of tropical deforestation and of biodepletion, both present and prospective. Even if all commercial logging, cattle ranching, fuelwood gathering, road building, dam construction, mining and other forms of deforestation were to be halted forthwith, well under half the problem overall would be solved. Yet we have scant knowledge and understanding of slash-and-burn cultivators. While their crucial role in deforestation was identified more than ten years ago (Myers, 1979; UNESCO, 1978), the amount of research directed at their activities has been all too little, and hardly any programmes of suitable scale and scope are envisaged.

N. Myers

Shifted cultivators

Today's slash-and-burn cultivators operate not so much as shifting cultivators. Rather, they constitute 'shifted' cultivators, i.e. displaced peasants who, finding themselves landless in established farming areas of the countries concerned, migrate to the last unoccupied public lands available where they can practice small-scale farming, i.e. tropical forests. Thus, they contrast strongly with the shifting cultivators of tradition, who made sustainable use of forestlands, as demonstrated by the fact that they practised their lifestyle for centuries without long-term damage to forest ecosystems.

Being powerless to resist the factors that impel them into the forests (see below), the shifted cultivators' lifestyle is driven by a host of factors – economic, social, legal, institutional, political – of which they have scant understanding and over which they have virtually no control (Myers, 1992; Westoby, 1989). So they are no more to be 'blamed' for deforestation than soldiers are to be held responsible for fighting a war. But they advance on the forest fringes in ever-growing numbers, pushing deeper into the forests year by year. Behind come more multitudes of displaced peasants, thus denying the forests any significant chance of regeneration (Manshard and Morgan, 1988; Palo and Salmi, 1988; Peters and Neuenschwander, 1988; Schuman and Partridge, 1989; Thiesenhusen, 1989; Geores and Bilsborrow, 1991; Bilsborrow and Hogan, 1994).

Driven significantly by population growth and sheer pressure on existing farmlands (albeit often cultivated with only low or medium levels of agrotechnology, hence cultivated in extensive rather than intensive fashion), slash-and-burn farming is the main source of deforestation in such prime tropical forest countries as Colombia, Ecuador, Peru, Bolivia, Ivory Coast, Nigeria, Zaire, Madagascar, India, Nepal, Sri Lanka, Thailand, Indonesia and the Philippines, often too in Mexico, Brazil, Myanmar and Vietnam (FAO, 1992; Grainger, 1993; IUSSP, 1992; Myers, 1989, 1992).

Moreover, the shifted cultivators' numbers are expanding fast, and the same for the amount of deforestation they cause. This contrasts with the other main agents of deforestation. Only one-fifth of current forest loss is due to over-heavy or otherwise excessive logging, which has hardly increased since the early 1980s. One-tenth is due to cattle ranching, which has actually been declining somewhat. One-seventh is due to road building, dam construction, commercial agriculture, fuelwood gathering and other peripheral activities, which have likewise shown scant increase. Almost

Table 10.1. *Tropical deforestation: main causes, 1989*

Deforestation activity	Expanse deforested (km^2) (and % of total deforestation)	
Destructive logging confined to SE Asia	30 000	(21)
Cattle ranching, confined to C. America and Amazonia	15 000	(11)
Dams, mining, road building	12 000	(8)
Plantations for tea, rubber, oilpalm, etc.	8000	(6)
Slash-and-burn agriculture	77 000	(54)
Total	142 000	(100)

Source: Myers (1989).

three-fifths is attributable to the slash-and-burn farmer (Table 10.1; Myers, 1989, 1992; FAO, 1992).

Moreover, population growth rates for shifted-cultivator communities are generally higher than those for national populations. In the Philippines, they have been increasing at a rate well over 4.0% per annum, whereas the national rate has been 2.4% (Cruz *et al.*, 1992). As we shall see shortly, in the Brazilian Amazonian state of Rondonia their numbers expanded at a remarkable 15% per annum during the late 1970s and early 1980s, whereas the rate for Brazil as a whole was less than 2.0% (Malingreau and Tucker, 1988). In a lengthy list of countries – for example, Ecuador (Bilsborrow, 1989), Peru (Dourojeanni, 1991), Bolivia (Larson, 1988), Madagascar (Ramamonjisoa, 1990), Myanmar (Lynch, 1991), Thailand (Feder *et al.*, 1988) and Indonesia (Poffenberger, 1990) – the picture is much the same, with a population growth rate for shifted cultivators often twice the national rate. For details, see Table 10.2, which gives an indication of the numbers of shifted cultivators who might well come on stream within the foreseeable future.

To this significant extent, population growth plainly plays a role in the build-up of shifted cultivators. Equally plainly, many other factors are at work, notably peasant poverty, maldistribution of farmlands, inequitable land use systems, lack of property rights and tenure regimes, low-level agrotechnologies, insufficient policy support for subsistence farming, deficient rural infrastructure and inadequate rural development generally (Colchester and Lohmann, 1992; Dorner and Thiesenhusen, 1991; Grainger, 1993; Repetto and Gillis, 1988; Southgate and Runge, 1990;

Table 10.2. *Population growth and socioeconomic factors in leading tropical forest countries*

Country	Population 1950 (millions)	Population 1993 (millions)	Population growth rate, 1993 (%)	Population in rural areas, 1993 (%)	Population projected 2000 (millions)	Population projected 2025 (millions)	Projected size of stationary population (millions)	Per-capita GNP, 1991 ($US)	Foreign debt 1991 (billions ($US)
Latin America									
Bolivia	3	8	2.7	49	9	14	22	650	4
Brazil	53	152	1.5	24	172	205	285	2920	117
Colombia	12	35	2.1	32	38	51	63	1280	17
Ecuador	3	10	2.5	45	13	17	25	1020	13
Mexico	27	90	2.3	29	99	138	182	2870	102
Peru	8	23	2.0	28	26	36	48	1020	21
Venezuela	5	21	2.6	16	23	33	41	2610	34
Asia									
India	362	897	2.1	74	1017	1380	1886	330	72
Indonesia	77	188	1.7	69	206	278	354	610	74
Malaysia	6	18	2.3	49	22	33	43	2490	21
Myanmar	18	44	1.9	76	51	70	96	n/a	5
Papua New Guinea	2	4	2.3	87	5	7	12	820	3
Philippines	20	65	2.5	57	74	101	140	740	32
Thailand	20	57	1.4	81	65	76	105	1580	36
Vietnam	24	72	2.2	80	82	107	159	n/a	n/a
Africa									
Cameroon	5	13	2.9	60	16	33	56	940	6
Congo	1	2	2.8	59	3	5	15	1120	5
Gabon	n/a	1	2.4	54	2	2	7	3780	4
Ivory Coast	3	13	3.5	60	17	38	67	690	19
Madagascar	5	13	3.3	76	15	34	49	210	4
Nigeria	41	95	3.1	84	128	246	382	290	35
Zaire	14	41	3.3	60	50	105	172	220	10

Note: n/a, not available.
Source: World Bank (1992, 1993)

Westoby, 1989). While the populations of tropical forest countries expanded by amounts ranging from 15 to 36% during the 1980s, deforestation expanded by almost 90% (Myers, 1989, 1991). Malaysia has only one-tenth as many people as Indonesia, but has cleared 40% as much forest as Indonesia. While the amount of forest destroyed per inhabitant in India is less than 0.2 of a hectare, the corresponding figure is six times higher in both Ivory Coast and Colombia (World Bank, 1992). Vietnam, Peru and Papua New Guinea are each clearing forests at a rate of 3500 km² per year, but annual population increase is 1.4 million in Vietnam, 450 000 in Peru and 90 000 in Papua New Guinea (WRI, 1992).

On top of causative factors within the countries concerned, there are exogenous factors at work too, for example international debt. Tropical forest countries owe about two-thirds of international debt totalling $(US)1.3 trillion. Debt-burdened countries often feel inclined to overexploit their forest resources in order to generate foreign exchange. More significantly, debt means there is so much less opportunity for governments to assign funds to the subsistence agriculture sector, or to rural infrastructure, or to the many other measures that would help reduce poverty and landlessness among the peasantry. So strong is the linkage in certain instances that a $(US)5 billion reduction in a country's debt can led to a reduction of anywhere between 250 and 1000 (occasionally more) km² of annual deforestation (Kahn and McDonald, 1992; see also Gullison and Losos, 1993). Debt relief would be a potent factor in the case of for example Peru, with its $(US)21 billion debt and annual deforestation of 3500 km², and Philippines, with debt of $(US)32 billion and annual deforestation of 2700 km².

The case of Rondonia and Acre, Brazil

A salient instance of the shifted cultivator has occurred in the two Brazilian states of Rondonia and Acre in southern Amazonia, with a combined area of 400 000 km². Deforestation there has been proceeding with exceptional speed (Lisansky, 1990; Malingreau and Tucker, 1988; Setzer and Pereira, 1991). In the mid-1970s the Brazilian government started to sponsor smallholder settlements in Rondonia, and by the early 1980s in Acre as well. The population of Rondonia was only 111 000 in 1975, but thereafter it soared to well over one million by 1986, for an almost ten times increase in just 11 years. In 1975 only 1200 km² of the state's forests had been cleared, but by 1985 the amount had grown to almost 28 000 km². Many of the chief deforesters were small farmers and peasants, clearing land mainly to graze cattle. Rondonia's herd grew more than 30-fold between 1970 and 1988 (Hecht et al., 1988).

Agricultural settlement grew so pervasive in Rondonia and Acre that during the dry season of 1987 some 55000 km² of forest were burned (Myers, 1989). Fortunately the situation has recently been relieved through marked shifts in government policy, and in 1992 deforestation in the whole of Brazilian Amazonia totalled only a little over 10000 km² (Instituto Nacional de Pesquisas Espaciais, 1992).

The migratory surge into Rondonia and Acre has been due in major measure to population growth in Brazil. With 53 million people in 1950, Brazil today has 152 million, and the total is growing at a rate of 1.5% per year (the projected total in the year 2100 is almost 300 million people). This is not to ignore, however, the other factors that are involved in mass migration into Rondonia and Acre. One is the maldistribution of farmlands in the main agricultural territories of southern Brazil, leading to growing throngs of landless farmers. Five per cent of farmers possess 70% of farmlands, while 70% cultivate only 5% – a skewed situation that is growing more acute (Lisansky, 1990). Some of the larger holdings amount to thousands of hectares, too large for the owner to cultivate more than half, while the hard-scrabble peasant over the boundary fence has to make do with half a dozen hectares or less (Monbiot, 1991). To this extent a response to the deforestation problem lies not only with population planning throughout Brazil, but with agrarian reform for the agricultural sector in lands far outside Amazonia. Indeed all Brazil's present farmers, plus those likely to be added throughout the next century, could be accommodated in farmlands outside Amazonia, supposing that land redistribution were to be undertaken on a sufficient scale.

Brazilian deforestation is also aggravated by factors of poorly defined property rights and ownership regimes, plus the problem of deforestation-derived externality effects that are scarcely reflected at all by patterns and practices of forestland exploitation (Hecht and Cockburn, 1990). Probably most important of all is the question of inadequate policy emphasis on the part of the government toward improving agrotechnologies, credit supplies, marketing networks and rural infrastructure generally for the small-scale farmer in the traditional farmlands of southern Brazil. This deficiency is compounded by an overall policy orientation on the part of the government that tends to favour industry over agriculture, urban interests over rural concerns, capital-intensive over employment-creating activities, and ultimately the richer over the poorer (Brazil features the most skewed wealth-and-income ratio in the world) (Binswanger, 1987; Mahar, 1989).

The future outlook

There is vast scope for still larger throngs of shifted cultivators to accelerate deforestation in many parts of the humid tropics. Given the 'demographic momentum' built into population-growth processes in countries concerned, and even allowing for expanded family-planning programmes, population projections (Table 10.2) suggest that, in those countries where economies appear likely to remain primarily agrarian, there will surely be progressive pressures on remaining forests, extending for decades into the future. For instance, Ecuador's population is projected to increase from 10.3 million today to 25 million (143% greater) before it attains zero growth in about a century's time; Peru's from 22.9 million to 48 million (110%); Cameroon's from 12.8 million to 56 million (338%); Zaire's from 41.2 million to 172 million (317%); Madagascar's from 13.3 million to 49 million (268%); Myanmar's from 43.5 million to 96 million (121%); Vietnam's from 71.8 million to 159 million (121%); and Indonesia's from 187.6 million to 354 million (89%).

So, unless there is resolution of the landless-peasant phenomenon – a prospect that appears less than promising (Salmi, 1988) – it is difficult to see that much forest cover will remain in another 30 or 50 years' time in many if not most tropical forest countries.

Of course we must be careful not to overstate the case. In much of the island of New Guinea, population pressures are slight to date, and appear unlikely to become predominant within the foreseeable future. There are only eight million people in 827 000 km^2, and they are expected to total no more than 15–20 million in the year 2025. Much the same applies in the countries of the Guyana shield, i.e. Guyana, Suriname and French Guiana: 1 337 000 people in 468 000 km^2 today, projected to reach no more than two million at most in 2025. We can say the same for the three countries of the Zaire Basin, though only for the present: 45 million people in 2.9 million km^2, though projected to total 112 million (for a 151% increase) in 2025.

Moreover, there is vast scope for still larger throngs of shifted cultivators to accelerate deforestation in many parts of the humid tropics. Given the demographic momentum built into population-growth processes in the countries concerned, and even allowing for expanded family-planning programmes, population projections (Table 10.2) suggest that in those countries where economies appear likely to remain primarily agrarian, there will surely be progressive pressures on remaining forests, extending for decades into the future.

As a further measure of the scope for increasing streams of shifted cultivators to migrate into the forests in the foreseeable future, consider two other determining factors. First, alternative forms of livelihood for landless peasants are becoming more limited by unemployment problems. Developing countries as a whole need to generate 700 million jobs (or as many as all the jobs in the developed world today) during the next 20 years simply to accommodate new entrants into the labour market, let alone to relieve present unemployment and underemployment, which often amount to 30–40% of the work force. During the 1990s the number of developing-country jobs required is 30 million per year; the United States, with an economy larger than all the developing countries' combined, often has difficulty generating two million new jobs per year. In Brazil, 1.7 million new people seek jobs each year, over half of them failing to find enough employment to support themselves, and many of them joining the migratory surge toward Amazonia.

Second, there is a growing problem of farmland shortage in many if not most developing countries, where land provides the livelihood for around three-fifths of populations and where the great bulk of the most fertile and most accessible land has already been taken (Harrison, 1992; Myers, 1994). A full 200 million farmers, with families totalling around one billion people, have too little land for minimum subsistence requirements of food and fuel (El-Ghonemy, 1990). Many of these rural poor are increasingly encroaching onto low-potential lands, generally marginal areas and principally tropical forests, where they have no option but to overexploit environmental resource stocks in order to survive (World Bank, 1990; Oodit and Simonis, 1992). Already 60% of the world's 1.2 billion people in absolute poverty inhabit environmentally fragile areas where it is extremely difficult for them to gain a sustainable livelihood (Leonard, 1987). In Indonesia, 15% of the rural population was functionally landless in 1988, in Thailand the same, in Peru 19%, in Myanmar 22%, in Ecuador 23%, and in the Philippines 34% (Jazairy et al., 1992). Worse, farmland shortages continue to spread among the rural poor at rates between 3% and 5% a year, i.e. faster than population growth rates (Jazairy et al., 1992).

Conclusion

There seems little prospect of resolving the biodepletion problem without resolving the tropical deforestation problem. In turn, there we have little hope of slowing deforestation (let alone halting it) without getting on top of the problem of shifted cultivators – or rather the plethora of problems they represent. A man with machete and matchbox in hand is not a problem, he is a *symptom* of a whole complex of problems. Hence the would-be saviours of biodiversity need to direct greater attention to the *sources* of problems that drive shifted cultivators into the forest, i.e. population growth, peasant poverty, lack of agrarian reform, and inadequate rural development among a host of other problems, all of which lie far outside tropical forests.

Yet the shifted cultivator remains a forgotten figure, ignored by governments and international agencies alike. So far as one can discern, this is often because tropical forests are in effect used by many governments as 'safety valves' for throngs of landless and unemployed people who might otherwise cause political trouble. Let these destitutes migrate to the forests, officials tacitly urge; if they riot there, who is to hear them? Ironically, governments also decry them as law breakers; the migrants are not supposed to burn the forests. All the more reason to view them bureaucratically as non-persons.

This is presumably why hardly a single government has mounted an effort to census (as opposed to censor) the forestland communities, nor have either of the two international agencies concerned (or rather, unconcerned), the Food and Agriculture Organization and the United Nations Population Fund.

For all their depredations, the cultivators are far less culpable than are the commercial loggers, the cattle ranchers and others who fell the forests. Loggers could obtain all their timber from tree plantations established on already deforested lands. Ranchers could raise much more beef on existing ranches if they were induced to practise even a modicum of pasture management. Slash-and-burn farmers have no such ready recourse. Not only do they lack land elsewhere; they are generally short of the agronomic techniques to enable them to grow their subsistence food on permanent forestland plots rather than to keep pushing deeper into the forests as their erstwhile farm patches lose natural soil fertility. If they could be supplied with the means to practise sustainable agriculture, they would often leap at a chance to abandon their destructive lifestyle. After all, clearing new forest plots is distinctly hard labour.

If the migratory surge of shifted cultivators continues, there could be one billion of these forestland inhabitants within another two decades at most. To the extent that that prognosis is correct, the migrant phenomenon represents by far the biggest mass movement of people to have occurred in the history of humankind in so short a space of time; and the destruction of tropical forests represents by far the biggest land use change to have occurred in such a similarly short space of time. The phenomenon of the shifted cultivator remains almost entirely undocumented and disregarded.

References

Bilsborrow, R. E. (1989). *Agricultural Colonization and Ecological Deterioration in the Amazon Region of Ecuador.* Carolina Population Center, University of North Carolina, Chapel Hill, NC.

Bilsborrow, R. E. and Hogan, D., (eds.) (1994). *Population and Deforestation in the Humid Tropics.* Oxford University Press, New York.

Binswanger, H. (1987). *Fiscal and Legal Incentives With Environmental Effects on the Brazilian Amazon.* Agricultural and Rural Development Center, The World Bank, Washington, DC.

Colchester, M. and Lohmann, L. (eds.) (1992). *The Struggle for Land and the Fate of the Forests.* Zed Books, London.

Cruz, M. C., Meyer, C. A., Repetto, R. and Woodward, R. (1992). *Population Growth, Poverty, and Environmental Stress: Frontier Migration in the Philippines and Costa Rica.* World Resources Institute, Washington, DC.

Dorner, P. and Thiesenhusen, W. C. (1991). *Land Tenure and Deforestation: Interactions and Environmental Implications.* Department of Agricultural Economics, University of Wisconsin, Madison, WI.

Dourojeanni, M. J. (1991). *Amazonia Peruana: que hacer?* Fundaciñ Peruana para la Conservación de la Naturaleza, Lima.

El-Ghonemy, R. M. (1990). *The Political Economy of Rural Poverty: The Case for Land Reform.* Routledge, London.

Feder, G. *et al.* (1988). *Land Policies and Farm Productivity and Thailand.* John Hopkins University Press, Baltimore, MD.

FAO (Food and Agriculture Organization) (1992). *Third Interim Report on the State of Tropical Forests.* Forest Resources Assessment 1990 Project, Food and Agriculture Organization, Rome.

Geores, M. E. and Bilsborrow, R. E. (1991). *Deforestation and Internal Migration in Selected Developing Countries.* Carolina Population Center, University of North Carolina, Chapel Hill, NC.

Grainger, A. (1993). *Controlling Tropical Deforestation.* Earthscan Publications, London.

Gullison, R. E. and Losos, E. C. (1993). The role of foreign debt in deforestation in Latin America. *Conservation Biology,* 7, 140–7.

Harrison, P. (1992). *The Third Revolution: Environment, Population and a Sustainable World.* Tauris, London.

Hecht, S. and Cockburn, A. (1990). *The Fate of the Forest.* Penguin Books, London.

Hecht, S. B., Norgaard, R. and Possio, G. (1988). The economics of cattle ranching in eastern Amazon. *Interciencia,* 13, 233–40.

Instituto Nacional de Pesquisas Espaciais (1992). *Deforestation in Brazilian Amazonia.* Instituto Nacional de Pesquisas Espaciais, Sao José dos Campos, Brazil.

International Union for the Scientific Study of Population (1992). *Proceedings of Seminar on Population and Deforestation in the Humid Tropics, Campinas, Brazil, 30 November–3 December 1992.* International Union for the Scientific Study of Population, Liege.

Jazairy, I., Alamgir, M. and Panuccio, T. (1992). *The State of World Rural Poverty: An Inquiry into its Causes and Consequences.* Intermediate Technology Publications, London.

Kahn, J. R. and McDonald, J. A. (1992). *Third World Debt and Tropical Deforestation.* State University of New York, Binghamton, NY.

Larson, B. (1988). *Colonialism and Agrarian Transformation in Bolivia.* Princeton University Press, Princeton, NJ.

Leonard, H. J. (1987). *Natural Resources and Economic Development in Central America.* Transaction Books, Oxford, and New Brunswick, NJ.

Lisansky, J. (1990). *Migrants to Amazonia: Spontaneous Colonization in the Brazilian Frontier.* Westview Press, Boulder, CO.

Lynch, O. J. (1991). *Community-Based Tenurial Strategies for Promoting Forest Conservation and Development in South and Southeast Asia.* World Resources Institute, Washington, DC.

Mahar, D. (1989). *Government Policies and Deforestation in Brazil's Amazon Region.* The World Bank, Washington, DC.

Malingreau, J.-P. and Tucker, C. J. (1988). Large-scale deforestation in the southeastern Amazon basin of Brazil. *Ambio,* **17,** 49–55.

Manshard, W. and Morgan, W. B. (eds.) (1988). *Agricultural Expansion and Pioneer Settlements in the Humid Tropics.* United Nations University, Tokyo.

Monbiot, G. (1991). *Amazon Watershed: The New Environmental Investigation.* Michael Joseph, London.

Myers, N. (1979). *The Sinking Ark.* Pergamon Press, Oxford.

Myers, N. (1989). *Deforestation Rates in Tropical Forests and Their Climatic Implications.* Friends of the Earth, London.

Myers, N. (1991). The world's forests and human populations: the environmental interconnections. In K. Davis and M. S. Bernstam, eds. *Resources, Environment and Population: Present Knowledge, Future Options,* pp. 237–51. Oxford University Press, New York.

Myers, N. (1992). *The Primary Source: Tropical Forests and Our Future.* W. W. Norton, New York.

Myers, N. (1994). Tropical deforestation: rates and patterns. In D. Pearce and K. Brown, eds. *The Causes of Tropical Deforestation,* pp. 27–40. University College London Press, London.

Oodit, D. and Simonis, U. E. (1992). *Poverty: Environment and Development.* Wissenschaftzentrum, Berlin.

Palo, M. and Salmi, J. (eds.) (1988). *Deforestation or Development in the Third World?* Finnish Forest Research Institute, Helsinki.

Peters, W. J. and Neuenschwander, L. F. (1988). *Slash and Burn Farming in Third World Forests.* University of Idaho Press, Moscow, ID.

Poffenberger, M. (1990). Facilitating change in forest bureaucracies. In *Keepers of the Forest: Land Management Practices in Southeast Asia,* Kumarian Press, West Hartford, CN.

Ramamonjisoa, B. S. (1990). *Analyse de la Filière Bois Malagache.* Direction des Eaux et Forêts, Antananarivo, Madagascar.

Repetto, R. and Gillis, M. (1988). *Public Policy and the Misuse of Forest Resources.* Cambridge University Press, Cambridge.

Salmi, J. (1988). Land reform – a weapon against tropical deforestation? In M. Palo and J. Salmi, eds. *Deforestation or Development in the Third World,* vol. II, pp. 159–80. Division of Social Economics of Forestry, Finnish Forest Research Institute, Helsinki.

Schuman, D. and Partridge, W. L. (1989). *Human Ecology of Tropical Land Settlement in Latin America.* Westview Press, Boulder, CO.

Setzer, A. W. and Pereira, M. C. (1991). Amazonia biomass burning in 1987 and an estimate of their tropospheric emissions. *Ambio,* **20,** 19–22.

Southgate, D. and Runge, C. F. (1990). *The Institutional Origins of Deforestation in Latin America.* Department of Agricultural and Applied Economics, University of Minnesota, St Paul, MN.

Thiesenhusen, W. C. (ed.) (1989). *Searching for Agrarian Reform in Latin America.* Unwin Hyman, Boston, MA.

UNESCO (1978). *Tropical Forest Ecosystems: A State-of-Knowledge Report.* Natural Resources Research XIV, UNESCO, Paris.

Westoby, J. (1989). *Introduction to World Forestry.* Basil Blackwell, Oxford.

World Bank (1990). *World Development Report 1990: Poverty.* Oxford University Press, New York.

World Bank (1992). *World Development Report 1992: Development and the Environment.* Oxford University Press, New York.

World Bank (1993). *World Development Report 1993: Investing in Health.* Oxford University Press, New York.

WRI (World Resources Institute) (1992). *World Resources 1992–1993.* Oxford University Press, New York.

Part D

Diversity decline as consequence of development

11
Human population dynamics and biodiversity loss

FRASER D. M. SMITH, GRETCHEN C. DAILY
AND PAUL R. EHRLICH

Introduction

The accelerating loss of biological diversity is primarily attributable to human activities. The relevant policy problem for biodiversity preservation is how efficiently to allocate effort to abating the interrelated factors driving biodiversity loss. The size of the human population is a multiplier of the sum of per capita impacts on biodiversity loss (see e.g. Ehrlich and Holdren, 1971; Holdren and Ehrlich, 1974), raising the following critical question. To what extent does the preservation of biodiversity hinge upon a reduction of the growth rate or size of the human population? Could per capita impacts on biodiversity be sufficiently diminished to make the total size of the human population relatively inconsequential? The answer has dramatic implications for the nature of the socioeconomic policy required to protect biodiversity.

In this chapter, we identify the chief causes of biodiversity loss and evaluate the relative contribution of human population size to each. Our focus here is on biophysical – as opposed to socioeconomic or political – dimensions of the maintenance of biodiversity and human well-being. We begin with a brief outline of current knowledge regarding human population. We then describe the role of population in each threat to biodiversity. We conclude with a general characterisation of the social and economic policies required to preserve as much as possible of the world's current and future remaining biodiversity.

Current patterns and projected changes in biodiversity and human population size

The diversity of biological entities resides at a number of levels of organisation, and can be defined in relation to almost any higher level: for

125

example, the diversity of proteins within a cell, genes within a population, or species within an ecosystem. However, biodiversity loss is conventionally characterised in terms of local or global species extinction. There are two principal reasons for this. First, although one of biology's grand old debates is how to define a species objectively, the species is still the most easily discriminated collective unit of biological organisation. Second, the loss of species diversity *appears* more irreversible than the loss of other kinds of diversity even if, in reality, this is not so.

From a policy perspective, maximising the benefits conferred on humanity by biodiversity appears to hinge at least as much upon the preservation of population diversity as upon species diversity (Ehrlich and Daily, 1993a; Daily and Ehrlich, 1994). Lack of quantitative information on population diversity forces us, however, to describe the relationship between human population dynamics and biodiversity loss in terms of species extinctions.

The total species richness of the world is not known with any accuracy. Estimates range from 3–10 million species (May, 1990) to 30 million (Erwin, 1982) to 50–100 million (Wilson, 1992); about 1.7 million species have so far been described, of which slightly less than one million are insects, about 250 000 are higher plants and 4500 are mammals.

Ignorance of total species number translates into even greater ignorance about rates of species loss. About 500 animal extinctions have been recorded since 1600 (Smith *et al.*, 1993a), which does not greatly exceed the normal evolutionary turnover of species. Species' lifetimes vary, but the median lifetime in the fossil record is about two million years (Stanley, 1975). Given that there are of the order of two million species known to science, we should expect an average of about one species per year to become extinct. The majority of recorded species are animals, rather than plants or microorganisms, so 500 animal extinctions since 1600 suggests that extinctions are proceeding at roughly the normal rate. But the majority of these extinctions have taken place among vertebrates, which constitute a small fraction of the world's species. On this basis, there may have been many more extinctions among small-bodied organisms in larger groups, such as insects, that went unnoticed. Also, the rate of recorded extinctions has been increasing since 1600 (only partly due to better sampling), and estimates of future extinction rates that circumvent the problem of knowing total species numbers or, in some cases, the actual rate of past extinctions, indicate that the current extinction rate may be orders of magnitude greater than the geological average.

Wilson (1992) drew on the ecological theory of island biogeography (MacArthur and Wilson, 1967) to estimate extinction rates in tropical forests. This empirically derived theory predicts that the number of species

(S) on an island is a function of island area (A) as given by the simple relation:

$$S = cA^z.$$

The exponent z has been found to vary from 0.21 to 0.30 for many systems, with an average of about 0.25 (see Begon *et al.*, 1986). At the margin, a unit reduction or increase in the area of a habitat will correspond to 0.25 species being lost or gained. Wilson (1992) also noted that most species occur in tropical moist forest, a habitat type that is being destroyed at an annual rate of roughly 1–2%. Combining these observations, he calculated that 0.25–0.5% of tropical moist forest species are being lost per year, which, if representative of species loss globally, translates into about 100–200 years for the loss of 50% of all species.

Smith *et al.* (1993*b*) obtained the same result employing an independent approach using data on endangered species compiled by the International Union for the Conservation of Nature (IUCN), and show that 10–30% of species in well-studied groups (birds and mammals among animals, and gymnosperms and palms among plants) are listed as being threatened with extinction. The rates of additions of new species to IUCN lists, along with the rates at which species are listed in progressively more critical categories of endangerment, indicate that 50% of all species may be extinct within 200 years, if well-studied groups such as mammals are representative of less well-studied groups such as insects. A number of other studies, extrapolating past trends or using the species–area relation, have produced estimates of global species loss of between 1% and 11% per decade (for a review, see Reid, 1992). These rates produce times to 50% global species loss of between 70 and 700 years. May (1988) calculated that if 50% of species will be eliminated in 50–100 years' time then, given that in the geological past such a process would take of the order of 50–100 million years, extinction rates in the near future will be of the order of a million times the geological average.

Although all of these estimates rely partly on untested assumptions and shaky extrapolations, their close concurrence underlines the likelihood that species extinction rates are, and will continue to be for decades at least, many orders of magnitude above background rates. There is little doubt that a mass extinction is presently under way.

Long-range human population projections are more reliable than predicted extinction rates (and long-range economic projections) because they rely on fewer assumptions, the major ones concerning future fertility and mortality rates. If fertility rates remain high, the human population size will simply grow until death rates cause a catastrophic crash; even if the average fertility rate fell instantly and miraculously to 1.7 children per family, global population would still grow to 7.8 billion around 2050, assuming no rise in death rates. A medium scenario predicts 10 billion people by 2050, stabilising at about 11.6 billion after 2200 (United Nations Population Division, 1993). Taken together, these scenarios indicate that the biosphere will almost certainly be faced with supporting at least double the 1990 population of 5.2 billion people.

Coupled with the inevitable growth in human population size is a critical need to increase per capita consumption of food, energy and other resources in many parts of the world. At present, roughly 1–1.5 billion people obtain insufficient nutrition and energy from their diets to carry out normal activities (Kates *et al.*, 1988; UNFPA, 1994) and an estimated 10 million people have died of hunger or hunger-related disease annually in recent decades (Dumont and Rosier, 1969; WHO, 1987; WRI, 1987). Food production has been on the decline on a per capita basis in Africa since 1967, in Latin America since 1982 and globally since 1984 (FAO, 1956–89; UNFPA, 1994). The prospects for future increases in food production keeping pace with population growth are highly problematic (Daily and Ehrlich, 1990; Ehrlich *et al.*, 1993).

It is often argued that hunger is merely the result of maldistribution of food, or of the means to obtain it. An analysis by Kates *et al.* (1988) suggested otherwise. Assuming that recent world food production is equitably distributed, is not diverted to livestock, and that 40% of the harvest is wasted (based on data from the FAO; see discussion by Ehrlich *et al.* (1993, p. 4)), the world could feed around six billion on a vegetarian diet. A diet with 15% of its calories derived from animal sources could be supplied to around four billion people, and a diet characteristic of North America or Northern Europe, with 30% on calories derived from animal sources, could feed less than half of the world's 1993 population of 5.5 billion people. Satisfying the needs of 10 billion people will probably require a near-tripling of food production.

Increases in overall energy consumption are expected to be similarly dramatic. Even under the most optimistic scenarios, which assume great increases in energy efficiency, total electricity production will more than triple by 2050 and direct fuel use will increase by about a third (Johansson *et al.*, 1993). The World Commission on Environment and Development (WCED, 1987) projects a 5- to 10-fold increase in global economic activity by 2040.

In the light of these projected increases in pressure on the world's biophysical systems – of which biological diversity is an integral component

– the following sections examine the principal anthropogenic threats to biodiversity and elucidate the role of human population growth in each.

Principal threats to biodiversity

Recorded threats from known extinctions and endangerments

Many records of past extinctions document the likely, or actual, causes of the disappearance of species. These records are instructive for our understanding of what kinds of human activity are most threatening to biodiversity. For the most part, records of causes of extinction are restricted to animals; the causes of plant extinctions are poorly documented. Of 486 animal species whose extinction has been documented since 1600 by the World Conservation Monitoring Centre (WCMC, 1992), 80 have become extinct from hunting and trapping by humans for food, skin, sport, live trade, or as pests; 114 have become extinct from the effects of purposeful or accidental introductions of exotic species and diseases into new areas by humans; 98 have become extinct from habitat destruction; 5 others have become extinct from other minor or uncertain effects. This leaves 189 for which no cause of extinction is known.

Some species currently threatened with extinction are known to be threatened by particular activities. The best records are for birds (see Collar and Andrew, 1988), of which just over 1000 out of about 9500 species are currently under threat (Smith *et al.*, 1993*b*). Of these, 122 species are threatened by hunting and trapping for food, sport, feathers or the bird trade, 76 are threatened by the introduction of exotic species into birds' ranges, 322 species are threatened by habitat destruction and about 30 species are threatened by a variety of miscellaneous causes.

Clearly, the three primary causes of *known* extinctions and endangerments are overexploitation (such as hunting and trapping), introduction of exotic species and habitat destruction. Of course, these causes are linked: for example, extinctions from the introduction of exotics accompany, and are often facilitated by, habitat clearance; introductions can themselves be viewed as a form of habitat modification or destruction. Figure 11.1 shows the numbers of known historical extinctions from overexploitation, introduction of exotics and habitat destruction.

Possible threats to biodiversity from global change

At the global level, certain phenomena such as ozone depletion and enrichment of the atmosphere by carbon dioxide and other trace gases that

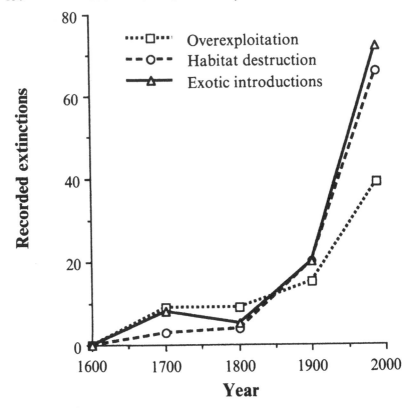

Figure 11.1. Non-cumulative graph of known historical species extinctions from overexploitation, habitat destruction and introduction of exotic species. Data are from WCMC 1992.

contribute to global warming may indirectly produce biodiversity losses. Unambiguous cases of extinction or endangerment from these phenomena cannot readily be established but studies linking ecological and physiological processes with the effects of anthropogenic global changes suggest how such extinctions might come about.

The depletion of the stratospheric ozone layer by anthropogenic emissions of chlorofluorocarbons (CFCs) and other compounds results in increased fluxes of ultraviolet radiation to the earth's surface in the waveband 280–320 nm (UV-B), especially at the poles. The extent of thinning of the ozone layer over Antarctica has increased since thinning was first detected (WRI, 1992). Holm-Hansen *et al.* (1993) have found that oceanic primary productivity in the Antarctic is reduced by less than 5% under current ozone thinning but ozone depletion is expected to increase well after CFC emissions have been halted because of the lag time of about 10 years from CFC emission to ozone depletion, and the 50–100 year persistence of CFCs in the stratosphere (WRI, 1992).

Increased UV-B fluxes reduce the rate of assimilation of carbon dioxide by land plants (Teramura *et al.*, 1990) but the sensitivity of plants to UV-B varies substantially among species. Sclerophyllous plants and conifers are relatively insensitive to increased irradiation with UV-B (Day *et al.*, 1992; Musil and Wand, 1993), whereas woody angiosperms and grasses show intermediate sensitivity, and herbaceous plants are very sensitive (Day *et al.*, 1992). This variation in response could result in shifts in the relative competitive abilities of species, with potential effects on the species composition of communities and on ecosystem processes (Kareiva *et al.*, 1993).

A general illustration of the problems of finding links between biodiversity losses and local or global anthropogenic effects is provided by declines in amphibian populations around the world. During the 1980s reports began to accumulate of amphibian species suddenly disappearing from places as far apart as California, Costa Rica, Brazil and Queensland (Weygoldt, 1989; Wake, 1991). These disappearances prompted concern among herpetologists for the survival of amphibians, which have been in existence for more than 100 million years, and have survived the mass extinction that terminated the dinosaurs. This concern led to a symposium on amphibian declines in 1990 (Wyman, 1990), at which evidence was presented that amphibians may be very sensitive to water and soil pollution (Wyman and Hawksley-Lescault, 1987), and atmospheric pollution. Amphibians may be acting as the global equivalents of a miner's canary so their worldwide declines should be a cause for concern outside – as well as within – the herpetological community, but so far the exact causes of the disappearances remain a mystery (Wake, 1991).

The rising concentrations in the atmosphere of carbon dioxide and other greenhouse gases are likely to produce global mean temperature rises of a few degrees centigrade over the next 100 years or so. This increase in carbon dioxide alone could change community species composition and ecosystem metabolism, affecting plant reproductive success, interspecific interactions and various processes of primary production in ecosystems (Houghton *et al.*, 1990). Furthermore, the magnitude and rate of change in local and regional climatic conditions is expected to exceed the capacity of many organisms to respond. In terrestrial systems, areas of anthropogenic disturbance will limit the capacity of species to migrate as climatic belts shift.

The geographical ranges of some species will respond faster than others

132   *F. D. M. Smith, G. C. Daily and P. Ehrlich*

to changes in climate. Rapid and substantial climate change might cause the geographical ranges of vagile species such as birds to move so far that ecological links would in many areas be broken: communities would be temporarily disassembled. Less vagile species such as plants may become marooned, and out-competed by invaders (see Murphy and Weiss, 1992; Root and Schneider, 1993). Disassembled communities would not necessarily return to their original states if and when stable vegetation patterns became re-established. Clearly, community disassembly puts some species at risk of extinction, such as those with restricted habitat or food requirements, those that are numerically rare, species that migrate, or species that are bound tightly into mutualistic interactions.

The role of human population size in threatening biodiversity

Having enumerated the main threats to biological diversity, we now examine the role of human population size in generating these threats.

Overexploitation (logging, firewood-gathering, fishing, hunting, trapping and trading)

Human population size clearly relates to the volume of logging, hunting, trapping and trading endangered species. Forests are being overexploited worldwide (Myers, 1989) and some tree species are doubtless being lost before they are even described by science. Many fish stocks have been overexploited, although no species is yet known to be forced to extinction. The trade in bird-of-paradise feathers reached a peak in the 1910s and 1920s, when the human population was less than two billion people (in this case, none of the bird species had been driven to extinction by the time a ban on feather-taking was imposed in 1928; see Gilliard, 1969). Because only certain animal species are hunted or traded, and these species – like most others – have localised distributions, only a relatively small proportion of the world's people have access to them.

Introduction of exotic species

Species extinctions and endangerment from the introduction of exotic species by human beings are indirectly linked with human population size. Throughout history and prehistory the migration of human populations brought non-human species into contact with each other, and this process has probably increased in rate with increasing human population size. The

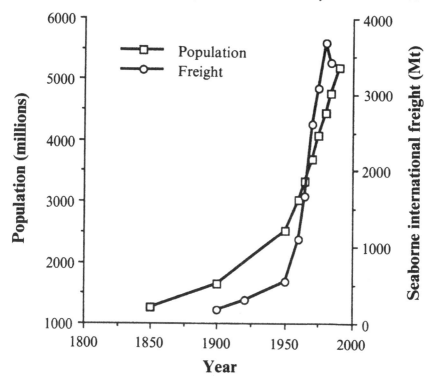

Figure 11.2. The increase in human population size since 1850 plotted against tonnage of seaborne international freight in megatonnes (Mt). Regression statistics: d.f. = 7; r^2 = 0.94; f = 87; p < 0.0001. Data are from Chisholm (1990) and Richards (1990).

rate of species extinctions from the introduction of exotics during the past few centuries correlates with human population size, which in turn correlates strongly with a measure of the volume of world trade (see Figures 11.1 and 11.2). All else being equal, more people trading more goods entails more contact between regions by land, sea and air. One of the most striking post-war examples is the decimation of the ground-dwelling birds of Guam by brown snakes (*Boigia irregularis*), which arrived there on supply ships during World War II (Rodda *et al.*, 1992). Brown snakes have recently been found on the runways at Honolulu International Airport, having stowed away in the wheel housings of aeroplanes – a danger signal for the native fauna of the Hawaiian islands, which include the greatest concentration of rare and endangered bird species in the world.

    The present rate of species introductions is also determined by the extent and severity of habitat disturbance and destruction. Undisturbed natural habitat is remarkably resistant to the establishment of exotic species, whereas disturbed areas are extremely susceptible to species introductions. This pattern is illustrated spectacularly by the contrast in species composition of Hawaiian lowland and upland forests, the former having experienced greater disturbance than the latter (see Groves and Burdon, 1986; Moulton and Pimm, 1986).

Habitat destruction

It is in the destruction of natural habitats by humans for agriculture, building materials and other purposes that we see the strongest links between biodiversity losses and human population increases (Ehrlich, 1994). In fact, species may be lost without outright clearing of vegetation because they may be sensitive to changes that are imperceptible to observers. The fragmentation of tracts of habitat into 'islands' separated by artificial vegetation may have devastating effects on biodiversity (see Forman *et al.*, 1976; Robbins, 1980); even low-intensity logging, or the alteration of the physical structure or age structure of vegetation, can have significant impacts (see Martin, 1992; Blankespoor, 1991).

The principal proximate cause of habitat destruction is the expansion of agricultural activities. Humanity already consumes directly, co-opts or destroys nearly 40% of global net primary productivity (NPP), the basic food supply of all animals (Vitousek *et al.*, 1986). Just as in water project development and mineral deposit exploitation, the most productive and accessible sources of NPP have been developed first. The less productive and accessible sources – often repositories of great biological diversity – are coming under rapidly increasing strain, as Figure 11.3 shows.

The preservation of biodioversity in the face of a needed tripling of agricultural output by the middle of the next century will not be possible without dramatic improvements in agricultural yields on land already converted to agricultural use. Achieving such improvements hinges upon making major changes in agricultural and economic policy worldwide, overcoming social obstacles at least as daunting as the biophysical constraints themselves (e.g. Daily, 1993).

Changes in global atmospheric conditions

Some anthropogenic alterations of global processes are more closely linked with human population growth than others. Although the effects of increased fluxes of ultraviolet light from ozone depletion may diminish

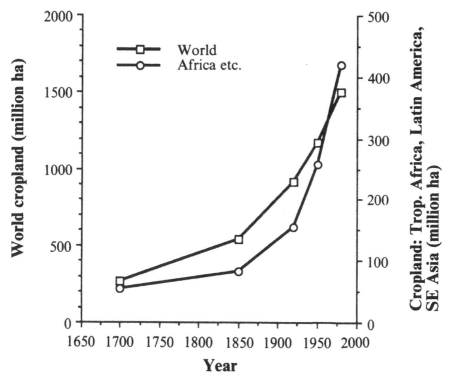

Figure 11.3. Increases in areas under cropland from 1700 to 1980, for the world (466% increase) and for tropical Africa, Latin America and Southeast Asia combined (762% increase). The latter group of areas holds the majority of the world's biological diversity. Data are from Richards (1990).

global plant diversity, these increases come largely from the activities of industrialised nations, and then only from certain industries. A rather small proportion of the world's population engages in activities that pollute the stratosphere with ozone-degrading CFCs. By contrast, the increasing emission of carbon dioxide into the atmosphere is more closely linked with global population size because a variety of human activities contribute to it. In less industrialised countries, increasing numbers of livestock and defor-estation contribute to the enrichment of the atmosphere with carbon dioxide, and in industrialised countries this enrichment – on a greater scale – occurs largely from factory and automobile emissions (WRI, 1992). Figure 11.4 shows the historical correlation between human population size and carbon dioxide concentration in the atmosphere. Although large-scale processes may appear to be indirectly linked with current or near-

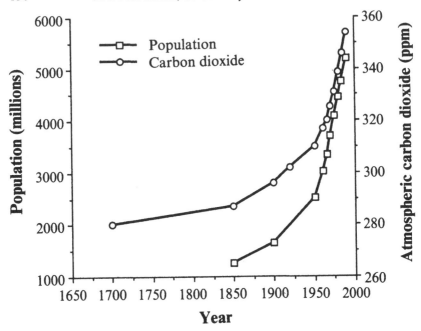

Figure 11.4. The increase in human population size since 1850 plotted against atmospheric carbon dioxide concentration in parts per million. Regression statistics: d.f. $= 8$; $r^2 = 0.99$; $F = 923$; $p < 0.0001$. Data are from Richards (1990) and US Congress, OTA (1991).

future biodiversity losses, and with human population increases, the apparent weakness of these links may equally be due to scientists' inability to measure them adequately (see also Ehrlich, 1994).

Policy implications

The kinds of human activity that threaten biodiversity are linked with human population growth to differing extents. Direct overexploitation of other species is a problem whose solutions can often be separated from solutions to human demographic problems. However, anthropogenic introductions of exotics into new areas – resulting in an incipient homogenisation of the biosphere – is rather more closely linked to human population dynamics through to the accelerating globalisation of trade. But the primary threat to biodiversity – especially natural population diversity – is the destruction of natural habitats, and this will become an overwhelmingly important factor in the near future, with the momentum of unprecedented human numbers and resource use behind it.

The long-term benefits of conserving biological diversity are widely understood and covered elsewhere (e.g. Swanson, 1992; WCMC, 1992). The life-support, economic, and existence values of biodiversity clearly make the conservation of biodiversity a critical policy goal, one that will become harder to ignore in the decades ahead. Can this policy goal be achieved by economic means alone, or are policies to halt and ultimately reverse human population growth also necessary?

At the meeting that spawned this volume, certain economic instruments were singled out for halting biodiversity loss. These include: the removal of subsidies distorting the economic benefits of agricultural practices that cause biodiversity loss; the establishment of equitable land tenure systems to draw people away from shifting 'frontier' cultivation, thus alleviating the pressure on natural habitats from slash-and-burn practices; and the intensification of agricultural production on lands already converted to agriculture, reducing the need to convert more land.

The first two economic goals are most urgent in the tropics, and the third is urgent in all parts of the world. The areas most suited to agriculture have already been converted, and much of this land is used inefficiently and unsustainably. Nevertheless, most species diversity is concentrated in the remaining areas suited to agriculture (see, e.g. Huston, 1993), and the gains in agricultural production that could result from converting these regions are very small compared with the potential gains in – indeed the necessity for, given the current global population momentum – increasing yields from lands in production now. However, agricultural intensification will require major changes in economic policy and farm practices to avoid concomitant intensification of environmental damage from agriculture (see NRC, 1989).

Another arena of economic policy with relevance to biodiversity loss is the distribution of consumption patterns. Rich countries exhibit unsustainable consumption in most resources, in that their consumption is supported by trade with the more biologically diverse poor countries, which in turn plunder their natural resources to supply this trade. The elimination of wasteful consumption in rich countries, the control of international debt, and the transfer of appropriate technologies and expertise from rich to poor countries will relieve pressure on natural habitats in poor countries.

Agricultural demands in the next few decades would by themselves increase per capita impacts on biodiversity but significant rises in such impacts will be driven mostly by increasing human population size. Even with major technological improvements in agriculture, the chances of success in conserving biodiversity will be significantly enhanced if economic instruments for this purpose are used in tandem with policies aimed at reducing fertility rates and improving other aspects of human well-being (see e.g. Ehrlich and Ehrlich, 1990, 1991; Dasgupta, 1993; Holl *et al.*, 1994). Put simply, the risk to humanity and other species of population collapse will be minimised if all reasonable means at humanity's disposal are used together to maximise the range of future economic options. As yet, there has been little attention given to how economic and demographic policies might best be combined to conserve natural systems and practice sustainable management. This is one arena where natural and social scientists can pool their skills to great benefit and, while the level of intellectual activity is promising, the distance yet to be covered remains daunting (see Ehrlich and Daily, 1993*b*; Daily, 1994).

Finally, there is the atmospheric commons, whose maintenance presents the largest challenge of all – literally and intellectually – to scientists and policy makers. Biodiversity preservation may ultimately hinge at least as heavily on sustainable management of the atmospheric commons as it does on protecting natural habitats, and management of the atmosphere clearly involves controlling human population size.

References

Blankespoor, G. W. (1991). Slash-and-burn shifting agriculture and bird communities in Liberia, West Africa. *Biological Conservation*, **57**, 41–72.

Chisholm, M. (1990). The increasing separation of production and consumption. In B. L. Turner, W. C. Clark and R. W. Kates, J. F. Richards, J. T. Mathews and W. B. Meyer, eds. *The World as Transformed by Human Action*, pp. 87–101. Cambridge University Press, Cambridge.

Collar, N. J. and Andrew, P. (1988). *Birds to Watch: ICBP Checklist of the World's Threatened Birds*. ICBP Technical Publication No. 8.

Daily, G. C. (1993). Social constraints on restoration ecology. In D. A. Saunders, R. J. Hobbs and P. R. Ehrlich, eds. *Nature Conservation III: Reconstruction of Fragmented Ecosystems*, pp. 9–16. Surrey Beatty and Sons, Perth.

Daily, G. C. (1994). Policy and philosophy for achieving environmental sustainability (Reviews of D. Pearce and J. Warford, *World Without End*, 1993; B. Cartledge, ed. *Energy and the Environment*, 1993; I. Simmons, *Interpreting Nature*, 1993). *Trends in Ecology and Evolution*, **9**, 155–6.

Daily, G. C. and Ehrlich, P. R. (1990). An exploratory model of the impact of rapid climate change on the world food situation. *Proc. Roy. Soc. Lond. B*, **241**, 232–44.

Daily, G. C. and Ehrlich, P. R. (1994). Population extinction and the biodiversity crisis. In C. Perrings, K. M. Mäler, C. Felke, C. F. Holling and B. O. Jansson, eds. *Biodiversity Conservation: Problems and Policies*, pp. 41–51. Kluwer Academic Press, Dordrecht.

Dasgupta, P. (1993). *An Inquiry into Well-Being and Destitution*. Clarendon Press, Oxford.

Day, T. A., Vogelmann, T. C. and Delucia, E. E. (1992). Are some plant life forms more effective than others in screening out ultraviolet-B radiation? *Oecologia*, **92**, 513–19.

Dumont, R. and Rosier, B. (1969). *The Hungry Future*. Frederick Praeger, New York.

Ehrlich, P. R. (1994). Energy use and biodiversity loss. *Phil. Trans. Roy. Soc. Lond.* B, **344**, 99–104.

Ehrlich, P. R. and Daily, G. C. (1993a). Population extinction and saving biodiversity. *Ambio*, **22**, 64–8.

Ehrlich, P. R. and Daily, G. C. (1993b). Science and the management of natural resources. *Ecological Applications*, **3**, 558–60.

Ehrlich, P. R. and Ehrlich, A. H. (1990). *The Population Explosion*. Random House, New York.

Ehrlich, P. R. and Ehrlich, A. H. (1991). *Healing the Planet*. Addison Wesley, New York.

Ehrlich, P. R., Ehrlich, A. H. and Daily, G. C. (1993). Food security, population and environment. *Population and Development Review*, **19**, 1–32.

Ehrlich, P. R. and Holdren, J. P. (1971). Impact of population growth. *Science*, **171**, 1212–17.

Erwin, T. L. (1982). Tropical forests: their richness in Coleoptera and other species. *Coleopterists Bulletin*, **36**, 74–82.

FAO (Food and Agriculture Organization) (1956–89). *Annual Production Yearbooks*, vols. 10–42. Rome.

Galli, A. E., Lek, E. S. and Forman, R. T. (1976). Avian distribution patterns in forest islands of different sizes in central New Jersey. *Auk*, **93**, 356–65.

Gilliard, E. T. (1969). *Birds of Paradise and Bower Birds*. The Natural History Press, Garden City, NY.

Groves, R. H. and Burdon, J. J. (eds.) (1986). *Ecology of Biological Invasions*. Cambridge University Press, Cambridge.

Holl, K. D., Daily, G. C. and Ehrlich, P. R. (1994). Causes and implications of the fertility decline and plateau in Costa Rica. *Environmental Conservation*, **20**, 317–23.

Holm-Hansen, O., Helbling, E. W. and Lubin, D. (1993). Ultraviolet radiation in Antarctica: inhibition of primary production. *Photochemistry and Photobiology*, **58**, 567–70.

Holdren, J. P. and Ehrlich, P. R. (1974). Human population and the global environment. *American Scientist*, **62**, 282–92.

Houghton, J. T., Jenkins, G. J. and Ephraums, J. J. (eds.) (1990). *Climate Change: The IPCC Scientific Assessment*. Cambridge University Press, Cambridge.

Huston, M. (1993). Biological diversity, soils and economics. *Science*, **262**, 1676–80.

Johansson, T. B., Kelly, H., Reddy, A. K. N. and Williams, R. H. (1993). *Renewable Energy*. Island Press, Washington, DC.

Kareiva, P. M., Kingsolver, J. C. and Huey, R. B. (1993). *Biotic Interactions and Global Change*. Sinauer, Sunderland, MA.

Kates, R. W., Chen, R. S., Downing, T. E., Casperson, J. X., Messer, E. and Millman, S. R. (1988). *The Hunger Report: 1988*. Alan Shawn Feinstein World Hunger Program, Brown University, Providence, RI.

MacArthur, R. H. & Wilson, E. O. (1967). *The Theory of Island Biogeography*. Princeton University Press, Princeton, NJ.

140   *F. D. M. Smith, G. C. Daily and P. Ehrlich*

Martin, W. H. (1992). Characteristics of old-growth mixed mesophytic forests. *Natural Areas Journal*, **12**, 127–35.

May, R. M. (1988). How many species are there on Earth? *Science*, **241**, 1441–9.

May, R. M. (1990). How many species? *Phil. Trans. R. Soc. Lond.* B, **330**, 293–304.

Moulton, M. P. and Pimm, S. L. (1986). Species introductions to Hawaii. In H. A. Mooney and J. A. Drake, eds. *Ecology of Biological Invasions of North America and Hawaii. Ecological Studies*, vol. 58, pp. 137–48. Springer-Verlag, New York.

Murphy, D. D. and Weiss, S. B. (1992). Effects of climate change on biological diversity in western North America: species losses and mechanisms. In R. L. Peters and T. E. Lovejoy, eds. *Global Warming and Biological Diversity*, pp. 355–68. Yale University Press, New Haven, CN.

Musil, C. F. and Wand, S. J. E. (1993). Responses of sclerophyllous Ericaceae to enhanced levels of ultraviolet-B radiation. *Environmental and Experimental Botany*, **33**, 233–42.

Myers, N. (1989). *Deforestation Rates in Tropical Forests and their Climatic Implications*. Friends of the Earth, London.

NRC (National Research Council) (1989). *Alternative Agriculture*. National Academy Press, Washington, DC.

Reid, W. V. (1992). How many species will there be? In T. C. Whitmore and J. A. Sayer, eds. *Tropical Deforestation and Species Extinction*, pp. 55–73. Chapman & Hall, London.

Richards, J. F. (1990). Land transformation. In B. L. Turner, ed. *The World as Transformed by Human Action*, pp. 163–78. Cambridge University Press, Cambridge.

Robbins, C. S. (1980). Effect of forest fragmentation on bird populations in the piedmont of the mid-Atlantic region. *American Naturalist*, **33**, 31–6.

Rodda, G. H., Fritts, T. H. and Conry, P. J. (1992). Origin and population growth of the brown tree snake, *Boigia irregularis*, on Guam. *Pacific Science*, **46**, 46–57.

Root, T. L. and Schneider, S. H. (1993). Can large-scale climatic models be linked with multiscale ecological studies? *Conservation Biology*, **7**, 256–70.

Smith, F. D. M., May, R. W., Pellew, R., Johnson, T. H. and Walter, K. R. (1993a). How much do we know about the current extinction rate? *Trends in Ecology and Evolution*, **8**, 375–8.

Smith, F. D. M., May, R. W., Pellew, R., Johnson, T. H. and Walter, K. R. (1993b). Estimating extinction rates. *Nature*, **364**, 494–6.

Stanley, S. M. (1975). A theory of evolution above the species level. *Proc. Natl Acad. Sci., USA*, **72**, 646–50.

Swanson, T. M. (1992). Economics of a biodiversity convention. *Ambio*, **21**, 250–7.

Teramura, A. H., Sullivan, J. H. and Ziska, L. H. (1990). Interaction of elevated UV-B radiation and carbon dioxide on productivity and photosynthetic characteristics in wheat, rice, and soybean. *Plant Physiology*, **94**, 470–5.

UNFPA (United Nations Population Fund) (1994). *Population, Food and Security*. Background document for the Meeting of Eminent Persons on Population and Development, Tokyo, Japan, 26–27 January 1994. Secretariat of International Conference on Population and Development, New York.

United Nations Population Division (1993). *Long Range World Population*

Projections: Two Centuries of Population Growth, 1950–2150. United Nations, New York.

US OTA (Congress OTA, Office of Technology Assessment) (1991). *Changing by Degrees: Steps to Reduce Greenhouse Gases.* OTA-O-482. US Government Printing Office, Washington, DC.

Vitousek, P. M., Ehrlich, P. R., Ehrlich, A. H. and Matson, P.A. (1986). Human appropriation of the products of photosynthesis. *BioScience*, **36**, 368–73.

Wake, D. B. (1991). Declining amphibian populations. *Science*, **253**, 860.

WCED (World Commission on Environment and Development (1987). *Our Common Future.* Oxford University Press, Oxford.

WCMC (World Conservation Monitoring Centre) (1992). *Global Biodiversity: Status of the Earth's Living Resources.* Chapman & Hall, London.

Weygoldt, P. (1989). Changes in the composition of mountain stream frog communities in the Atlantic Mountains of Brazil: frogs as indicators of environmental deteriorations? *Studies on Neotropical Fauna and Environment*, **24**, 249–56.

WHO (World Health Organization) (1987). *International Health News*, September 1987.

Wilson, E. O. (1992). *The Diversity of Life.* Belknap, Cambridge, MA.

WRI (World Resources Institute) (1987). *World Resources 1987.* Basic Books, New York.

WRI (World Resources Institute) (1992). *World Resources 1992–1993.* Oxford University Press, New York.

Wyman, R. L. (1990). What's happening to the amphibians? *Conservation Biology*, **4**, 350–2.

Wyman, R. L. and Hawksley-Lescault, D. S. (1987). Soil acidity affects distribution, behaviour and physiology of the salamander *Plethodon cinereus. Ecology*, **68**, 1819–27.

12

Scale and the feedback mechanism in market economics

COLIN W. CLARK

Recently a noted ecologist lost a bet to an equally noted economist. The ecologist had predicted that by the year 1990 the world economy would be affected by severe shortages of basic resources, with concomitant disruption of resource markets. When the predicted events failed to materialize, the bet was lost.

The same ecologist has also long been forecasting severe environmental deterioration. It might perhaps be argued that, if anything, actual environmental degradation has exceeded the worst of such prognostications. Neither acid rain, nor destruction of the ozone layer, to mention two notorious examples, featured prominently in the forecasts of environmental Doomsdayists a quarter of a century ago. The loss of biodiversity is another environmental concern that has only recently gained widespread notice.

Predicting the future is a risky business at best, particularly where human activities are involved. Humans, unlike any other species, have the ability to respond to threats to their common well-being through technological innovation and institutional organization. A person of optimistic bent might therefore predict a rosy future, with dynamic developments of a uniquely human character perpetually countering economic forces that tend to deplete and despoil the world's stock of productive and environmental resources. Even continued growth of the human population might not be seen as a subject of concern to such an optimist.

A closer examination, however, reveals a sharp contrast between the economics of global resource production and that of global environmental deterioration. I illustrate this dichotomy with two highly stylized systems models, (Figures 12.1 and 12.2).

For the case of resources, the global economy seems to work in the following way (Figure 12.1). As the economy uses up resources, proven

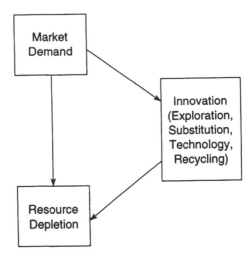

Figure 12.1. The World Resource System.

reserves are indeed depleted; individual stocks of some resources may cease production entirely. But the growing worldwide shortage – or even the threatened shortage – of a given resource generates market signals that render numerous innovations profitable. Entrepreneurs are motivated to seek new reserves, or to discover more cost-effective means of recovering the resource from existing reserves. Processes are developed that use the resource more efficiently, or make increasing use of substitute resources. All these innovations are market-driven; government intervention is entirely unnecessary.

In a sense, the industrial economic system rides along the edge of a resource precipice, always being in apparent imminent danger of resource depletion, but simultaneously generating automatic feedback signals that continually act so as to renew the resource base.

This is not to say that the market-driven feedback system is destined to succeed indefinitely. History has seen the rise and fall of numerous civilizations. The growth of these early civilizations followed technological breakthroughs, such as crop domestication, irrigation, and transportation, which initially facilitated the expansion of populations and economies. Eventually, however, the expansion of the population led to the depletion of resource stocks, or to environmental changes (soil loss, salinization, deforestation) that severely reduced agricultural productivity. Sooner or later these developments outpaced technological innovation. Combined with inevitable political and institutional failures, they ultimately led to collapse of the civilization.

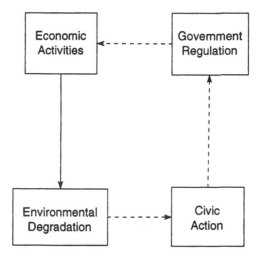

Figure 12.2. The World Environmental System.

Nothing inherent to the feedback loop of Figure 12.1 inhibits the expansion of scale of the entire system. It thus becomes a matter of faith to believe that the system will continue to operate, on an ever-expanding scale, indefinitely. Until now continued innovation has succeeded in keeping contemporaneous industrial civilization well away from the precipice. So successful has this global resource feedback system been that few people in the free, developed sector of the globe foresee any difficulties with resource supplies. Those few that have predicted disastrous resource shortages have repeatedly been proved wrong. Because of the dynamics of the global resource system, it seems unlikely that critical resource shortages are imminent today. Of course, particular resource stocks may be badly managed (as in the case of many marine fisheries, for example), and certain local human populations (such as those in the Sahel) may suffer from local resource shortages. But the overall picture of resource dynamics shown in Figure 12.1 still seems relevant on the global scale.

Resource extraction, however, is only one side of the economy–environment interaction. Equally significant is the generation of waste products and their ultimate disposal into the environment. Often these waste products are far from innocuous, consisting of materials not previously experienced by life forms, at least in the concentrations produced by industrial activities. These toxic wastes can directly affect human health (a significant portion of cancers are now thought to be induced by environmental pollutants), or can affect the economic system by interfering with important ecosystem processes. Indeed, with the prospect of global warming we see that waste products (carbon dioxide, methane) may ultimately affect global physical systems, with possibly extreme economic consequences.

The global environmental system (Figure 12.2), unlike the global resource system, lacks any automatic self-correcting feedback loop. If anything, market incentives act to intensify and perpetuate environmental pollution. Disposing of wastes in an environmentally benign way is costly, and any firm that can evade or minimize these costs will be at a competitive advantage. To the extent that there is any feedback loop to react to environmental degradation, it is generally a weak and indirect loop, as indicated in Figure 12.2. Environmentalists may (note: *may*) pressure the government to act; the government may enact antipollution legislation, which may cause polluters to reduce their toxic waste outputs. Meanwhile the polluting firms will be doing everything in their power to stem this process. They will lobby for weak legislation, for delays and exceptions in enforcement, for minimal penalties, and so forth.

And it is not just profit-oriented firms that behave in this way. Municipal governments, not wishing to increase taxes, pollute lands and waters with garbage and minimally treated sewage. The military, not wishing to expend its budget on waste clean-up, has now become a major polluter in many countries. Even individual citizens try to evade the costs of maintaining a clean environment; they drive polluting cars, light polluting fires, throw trash along roads, and worse. A few public-minded people, and possibly an occasional public-minded firm, may behave more responsibly, but the bottom line remains the same: polluting is cheap, cleaning up is expensive.

In cases where the potential consequences of environmental destruction are clear-cut, government action to limit emissions is possible. The Montreal Protocol regarding the phasing out of ozone-depleting compounds is a notable case in point. Yet it is estimated that a reversal of the present thinning of the ozone layer will not be possible for over 100 years. The economic consequences in the meanwhile are anyone's guess. Possible effects of ozone depletion are: decreased rates of photosynthesis, weakening of immune systems in plants and animals, and increased incidence of human and other cancers. The fact that the expansion of the polar ozone 'holes' has, year after year, far exceeded the most pessimistic forecasts is hardly a cause for indifference.

The two global systems – resource and environmental – are coupled in several important ways. The waste products that pollute the environment are merely transformed resources that are no longer useful. The larger the global resource system, the greater the resulting environmental impact.

As resources become scarcer or more expensive to extract, increasing amounts of used resources, formerly treated as waste, will be recycled. The second law of thermodynamics, however, limits the degree to which high-entropy wastes can be tranformed into valuable low-entropy resources. Technology, by its very nature, is characterized by the manufacture of increasingly complex products containing relatively small amounts of many different kinds of raw material (e.g. computers contain infinitesimal amounts of gold). The more complex the product, the more useless it becomes as waste. Furthermore, the progress of technology generates additional waste simply by rendering otherwise useful products obsolete. For example, building-demolition wastes have now become major (and highly polluting) components of landfill sites.

The global resource and environmental systems are also coupled in another way. As easily produced stocks of resources are used up, resource production itself begins to generate increasing levels of pollution. Billions of yards of rock are pulverized annually to extract increasingly diffuse minerals. The leachate from the finely crushed slag often contains toxic substances that poison water sources. The rock-crushing process uses increasing amounts of fossil fuels, adding to the flow of carbon dioxide into the atmosphere. In other words, the feedback loop operating in the resource system often acts so as to intensify the severity of environmental pollution.

Ultimately, it might be argued, some environmental feedback loop must come into being. When the environment becomes so polluted that costs of production are affected, surely the polluters will wake up to the need for environmental clean-up. This argument might be valid if polluters always had to live in their own excretions. But the very purpose of smoke stacks and discharge pipes is to ensure that this never happens. How many of today's factories could survive if 100% of their wastes were internalized?

The question of scale of the two systems again arises. As we have seen, nothing within these systems acts to limit their expansion of scale. Yet our planet is finite. It is especially ironic to think that, after centuries of worrying about resource depletion – an activity participated in by some very distinguished scholars (economists as well as biologists) – the most critical limit to growth may turn out not to lie after all in the global resource system, but in the planet's ultimate capacity for absorbing the wastes of our exponentially exploding economic system.

Every human being on earth, as well as every organization, adds daily to this ever-expanding flow of environmental pollutants. The very success of the global resource system results in a progressive expansion in scale of the global environmental problem. Individual rationality, in complete contrast to the case of resource shortages, encourages rather than discourages degradation of the environment. Effective negative feedback occurs only through social action, which means government, politicians, and corruption. No previous civilization has survived this dilemma, and there is as yet little evidence that our present industrial civilization will succeed in doing so.

13

Can economics protect biodiversity?

JONATHAN ROUGHGARDEN

I offer some comments on what economists need to learn about ecology.

Ecological knowledge

An ecologist earns a living by finding out how ecological systems work. To determine the state of a system, such as how many species are in it, is to describe the system. To discover the processes (or mechanisms) that bring about its state is to explain the system. Both the description and explanation of ecological systems is our job, and we have made great progress recently, although much remains unknown.

Now, the loss of biodiversity poses a problem. As citizens, it is our responsibility to call attention to the loss of biodiversity, just as it is our responsibility to report a building on fire. The policy question then is whether to let the fire burn, or to try to put it out. The policy objective is to protect human life and property. The policy implementation is to buy some water to put on the fire to save some of the building. An engineer, though not a physicist, might tell us how much of the house each additional gallon of water may save, and an economist might calculate how much water to buy. But this comfortable relationship among disciplines is lacking between ecologists and economists. We have no ecological engineers to say how an ecosystem will change if we choose to eliminate its species, one by one. That is not a question we have been trying to answer – we have been trying to find out what the species actually *in* the system are doing. So, there will be a time lag as information is developed on how particular systems will function if some of their components are eliminated.

Still, we do know several general facts. We know that ecosystems provide services whose loss will be expensive to replace, but that we do not buy through market mechanisms – we do not buy our day's supply of clean air

149

from the owner of a neighboring forest and so we can not turn to markets to measure the value of such ecosystem services. We also know that only some of the species in an ecosystem at any one time are important in providing the services that humans want, but that others may assume those roles when conditions change, particularly when epidemics damage the species that are presently most important. We also know that the aesthetic and recreational value of natural habitat is lost when enough of its components are removed, and we know that some species have a *Mona Lisa*-like value because they are very old, very curious, or somehow else very special. We also know that tropical rain forests and coral reefs especially are the repositories of evolutionarily tested chemicals that serve as drugs, spices, dyes and perfumes. It is obvious that general welfare tends to be hurt by the loss of biodiversity, and stewardship on its behalf is a public objective. Thus, the mandate to prevent the loss of biodiversity is obvious. In general terms, the loss of biodiversity is caused by habitat loss – if you want to kill a species, you kill its habitat. Therefore, general policies that prevent habitat destruction will, on the whole, ameliorate the decline of biodiversity. This is the *raison d'être* of this conference, and I am therefore enthusiastic about its aims.

Even if ecologists knew more details about how particular ecosystems function, and knew how those functions would change with human modification, economists would not know what to do with this information. How to value biodiversity is a research topic in ecological economics because it involves issues of common goods, portfolio management in uncertain environments, insurance premiums based on the public's preference for risk aversion, and the price of innovation. Economists must be very careful not to assume that biodiversity has little value simply because they have not yet succeeded in understanding how to find out what its value is. They should not assume that ecologists have some burden of proof to show that biodiversity is valuable, while allowing environmental destruction to proceed until such proof is at hand. To the contrary, economists should urge caution in environmental development until they can successfully value biodiversity taking into account the great many facts that are already known.

Economics is not morality

Although the economic theory of capitalism may qualify as a political philosophy, it is not a moral philosophy. Yet phrases such as 'we need to take costs into account when setting environmental objectives' confuse economics with morality. In fact, we *should not* take costs into account in setting environmental (or other) objectives, but we should take costs into account when considering how to implement moral objectives as policies. The cons, pros and imperatives of moral reasoning are not the same as the costs, benefits and axioms of economics. And if an economist were to argue that morality could be embedded in economics, serious discussion would be needed about whether a market mechanism can separate right from wrong. Therefore, it is necessary to reassert the moral objective of stewardship – our posterity should inherit a healthy social and natural environment.

Key ecological facts

Economists should know more about the fundamental causes of biodiversity loss than they do. These fundamental causes are not, of course, social causes. The causes of biodiversity loss are mechanisms that cause death rates in natural populations to exceed birth rates, and humans (or Martians for that matter) could activate such mechanisms. It happens that there is a statistical relationship between the number of species in a habitat and the size of the habitat, a result particularly clear on islands where a power law relates the number of species to island area. There is also a statistical relation between the number of species on an island and its distance from a source fauna. In birds and insects the number of species on an island of a given size and distance from a source fauna is a steady state resulting from the immigration rate of new species to the island coupled with the extinction rate of those already on the island. In this 'turnover' account of island biogeography, the species on an island are a random sample of the source fauna. Islands with two bird species, for example, have random pairs drawn from the urn of all bird species in the source fauna. In reptiles and mammals, there is also a statistical relation between island area and number of species. The processes determining the identity of the species are quite different from those for birds and insects. On islands that have relatively recently been formed by sea level rise, representing fragments of a formerly large island, a 'nested subset' distribution is seen. The smallest islands have one species, say A, and the next larger class of islands have A and B, then the next larger class A, B, and C. Thus, each island's fauna is a proper subset of a larger island's fauna. Conservation practice must be quite different for turnover versus nested-subset islands. Many small turnover-islands are equivalent to one large turnover-island, but many small nested-subset islands are identical with each other and do not conserve the species found only on large nested-subset islands. Finally, the fauna on many islands is determined by plate-tectonic motions, and dispersal of foreign species threatens biodiversity by tending to homogenize the world's biota. Ecologists do not know in general what the connection is between habitat area and the birth and death rates that ultimately determine whether a species becomes extinct, although this topic is receiving active research and may be solved shortly.

Discussions of the relation between biodiversity and habitat area omit reference to the value of the species for such purposes as sources of drugs, spices and so forth, as though all species are equal in their potential. Instead, species in highly diverse ecological communities tend to have more bioactive secondary compounds than do species from simple communities. This has been explained by saying that species in the tropics have to adapt to one another while species in the temperate zone have to adapt to the physical environment. If habitat reduction causes a reduction in the number of species, then the pressures of natural selection to produce bioactive chemicals drops, and species will slowly lose them and make, say, more leaves instead. The speed by which the organisms will evolutionarily lose their bioactive chemicals might be quite fast, say 100 generations, which for many populations is about 100 years. Also, if the synthesis of bioactive chemicals is physiologically induced, their loss could take place much faster.

Ecologists have also analyzed the spatial distribution of diversity. In a region with, say, 100 species, all might coexist throughout the region, or alternatively, each may be found all by itself in 100 different places. These are called the 'within-habitat' and 'between-habitat' components of a region's diversity, and the allocation to these components varies among the continents. If the regional biodiversity is mostly between-habitat diversity, then loss of a 'marginal hectare' has more impact than if the biodiversity is mostly within-habitat.

Conserving biodiversity requires some sensitivity to how species interact with each other. An ecological community is neither a house of cards that falls apart when one species becomes extinct, nor a bag of independent particles. Communities have many keystone species in them, species whose removal leads to the loss of further species. Predators, pollinators, tree-hole borers, decomposers and so forth can all have a role in which other species depend on them, either for creating a resource, providing access to a resource, or ameliorating the impact of a predator or superior competitor. Perhaps one species in ten is a keystone species. In ecological jargon, a niche is a species' occupation (gleaning insects from the undersides of leaves) and its habitat is its address (a swamp in Florida). A collection of interdependent species form the ecological analogue of an industry, and the removal of a keystone species removes an ecological industry from within the community. Also, the occupations of many species do not keep them in one place. Many species depend on the recurrence of gaps in the forest from tree falls, fire or hurricanes for their regeneration. Reconfiguring the geometry of a habitat affects the traveling-salesman types of species and can put them out of business.

The dynamics of marine and terrestrial communities appear to be quite different because marine organisms usually have a larval phase that lives in the water and disperses large distances. This couples the dynamics of distant sites to one another more than on land. Organisms do not live in air – they simply fly through it briefly, whereas organisms do live in water as they feed and reproduce. The consequences of this for biodiversity are not well understood, but it appears that marine species tend to be less endangered than are terrestrial species, and that marine habitats can recover from damage more readily than terrestrial habitats.

Economists have abstracted the problem of biodiversity loss to a degree that seems absurd to an ecologist. Of course, I am happy to support macroeconomic policies that retard land conversion from its natural state to some degraded state, but it is clear that the present level of economic analysis will barely be relevant to specific provinces or regions where a serious question of 'cut it down' or 'leave it alone' may be posed. At this scale, policy will have to incorporate some ecological science.

Can economics protect biodiversity?

So, we come to the bottom line, should ecologists collaborate in developing an economic approach to environmental preservation? In my opinion three issues have to be solved before we should trust economics with protecting the environment, including its biodiversity.

First, economic theory must be fundamentally improved to take account of the dynamic changes in the environment that different policies make. The basic approach of environmental economics to biodiversity is to imagine that some curve for the marginal benefit of biodiversity is plotted against a curve for the marginal cost of biodiversity, and the intercept is offered as the economically best level of biodiversity. Then externalities are internalized with taxes or tradable permits, and the 'band plays on'. The primitiveness of how economists value biodiversity has been mentioned above, and immediately biases this type of analysis. But, more importantly, the ecological feedbacks are not included. This is not to be confused with attacks on general equilibrium theory or calls for dynamic optimization. Instead, I want to see the chain rule of calculus used on environmental variables as well as social variables. If one adds a tax, then the impact on earnings, rents, labor and so forth are all calculated. But there is also an impact on the environment too that should be calculated. One may think of this point another way. The budget constraint is fundamental to 'economizing', the basic objective of economic theory. There is also an environmental constraint, and this should be added as a general axiom to equilibrium theory to supplement the budget constraint. Then the chain rule can be used with this constraint as well as with the budget constraint to obtain the policy implications for environmental variables as well as social variables. Furthermore, when dynamic optimization models are used, the equations that deal with the natural environment should contain expressions that ecologists have actually worked with, not thinly disguised economic myths such as the Cobb–Douglas formula.

Second, economists have to respect consumer sovereignty concerning environmental goods and services. Economists seem to assume that everyone would really rather have another automobile than protect a salamander. But as one comes to know cars and salamanders, it is a more difficult choice. Cars are not all that much fun, and it is easy to develop a respect for salamanders (and other species), not unlike forming a relationship with a pet. Ecologists have experienced this better than the general public because they have dealt with more species in the course of their work than most people have. Environmentalists are just as entitled to advertise information about the attractiveness of environmental goods and services as manufacturers are about cars. Economists should stay neutral about consumer preferences for environmental products, just as they are about other goods and services.

Third, economists have to distance themselves from their irresponsible fringe. Economists are continually confusing ecologists with environmental advocates. Economists often ask ecologists to be 'reasonable' when disputes over environmental protection arise. In fact, the Ecological Society of America has committed its support to a Sustainable Biosphere. Economists do not realize the significance of this. As a matter of record, the major professional society of ecologists in the United States is *for* development, provided it is sustainable. In contrast, economists look the other way when their brethren attack ecologists, environmentalists, and anyone slightly green, in newspaper column after newspaper column. They do not complain about the biased and uninformed comments of the *Wall Street*

Journal or the *Economist*. In private, economists disdain the column writers
– they are not real economists, just showmen. I think that before we trust
economists with the environment, they should publicly affirm the principle
of stewardship.

Index

Figures and Tables are indicated by *italic page numbers*, footnotes by suffix 'n'.

157

Printed in the United States
By Bookmasters